Asteroid Belt History and Terrestrial Bombardment

Luis D. Crowley

TABLE OF CONTENTS

TABLE OF CONTENTS – *Continued*

ABSTRACT

The main asteroid belt spans $\sim$ 2–4 AU in heliocentric distance and is sparsely populated by rocky debris. The dynamical structure of the main belt records clues to past events in solar system history. Evidence from the structure of the Kuiper belt, an icy debris belt beyond Neptune, suggests that the giant planets were born in a more compact configuration and later experienced planetesimal-driven planet migration. Giant planet migration caused both mean motion and secular resonances to sweep across the main asteroid belt, raising the eccentricity of asteroids into planet-crossing orbits and depleting the belt. I show that the present-day semimajor axis and eccentricity distributions of large main belt asteroids are consistent with excitation and depletion due to resonance sweeping during the epoch of giant planet migration. I also use an analytical model of the sweeping of the ν_6 secular resonance, to set limits on the migration speed of Saturn.

After planet migration, dynamical chaos became the dominant loss mechanism for asteroids with diameters $D \gtrsim 10$ km in the current asteroid belt. I find that the dynamical loss history of test particles from this region is well described with a logarithmic decay law. My model suggests that the rate of impacts from large asteroids may have declined by a factor of three over the last ~ 3 Gy, and that the present-day impact flux of $D > 10$ km objects on the terrestrial planets is roughly an order of magnitude less than estimates used in crater chronologies and impact hazard risk assessments.

Finally, I have quantified the change in the solar wind ^{6}Li/^{7}Li ratio due to the estimated in-fall of chondritic material and enhanced dust production during the epoch of planetesimal-driven giant planet migration. The solar photosphere is currently highly depleted in lithium relative to chondrites, and ^{6}Li is expected to be far less abundant in the sun than ^{7}Li due to the different nuclear reaction rates of the two isotopes. Evidence for a short-lived impact cataclysm that affected the entire inner solar system may be found in the composition of implanted solar wind particles in lunar regolith.

CHAPTER 1

DYNAMICAL HISTORY OF THE MAIN ASTEROID BELT

1.1   Introduction

The main asteroid belt spans the $\sim$ 2–4 AU heliocentric distance zone that is sparsely populated with rocky planetesimal debris. Strong mean motion resonances with Jupiter in several locations in the main belt cause asteroids to follow chaotic orbits, cross the orbits of the major planets, and be removed from the main belt (**Wisdom, 1987**). These regions are therefore emptied of asteroids over the age of the solar system, forming the well-known Kirkwood gaps (**Kirkwood, 1867**). In addition to the well known low-order mean motion resonances with Jupiter that form the Kirkwood gaps, there are numerous weak resonances that cause long term orbital chaos and transport asteroids out of the main belt (**Morbidelli and Nesvorny, 1999**). A very powerful secular resonance that occurs where the pericenter precession rate of an asteroid is nearly the same as that of one of the solar system's eigenfrequencies, the ν_6 secular resonance, lies at the inner edge of the main belt (**Williams and Faulkner, 1981**).

The many resonances found throughout the main asteroid belt are largely responsible for maintaining the Near Earth Asteroid (NEA) population. Non-gravitational forces, such as the Yarkovsky effect, cause asteroids to drift in semimajor axis into chaotic resonances whence they can be lost from the main belt (**Öpik, 1951; Vokrouhlický and Farinella, 2000; Farinella et al., 1998; Bottke et al., 2000**). The Yarkovsky effect is size-dependent, and therefore smaller asteroids are more mobile and are lost from the main belt more readily than larger ones. Asteroids with $D \lesssim 10$ km have also undergone appreciable collisional evolution over the age of the solar system (**O'Brien and Greenberg, 2005; Cheng, 2004; Bottke et al., 2005a**), and collisional events can also inject fragments into chaotic resonances (**Wetherill, 1977;**

Gladman et al., 1997). These processes (collisional fragmentation and semimajor axis drift followed by injection into resonances) have contributed to a quasi steady-state flux of small asteroids ($D \lesssim 10$ km) into the terrestrial planet region and are responsible for delivering the majority of terrestrial planet impactors over the last ~ 3.5 Gy (Bottke et al., 2000, 2002a,b; Strom et al., 2005).

In contrast, most members of the population of $D \gtrsim 30$ km asteroids have existed relatively unchanged, both physically and in orbital properties, since the time when the current dynamical architecture of the main asteroid belt was established: the Yarkovsky drift is negligble and the mean collisional breakup time is > 4 Gy for $D \gtrsim 30$ km asteroids. Asteroids with diameters between $\sim 10\text{--}30$ km have been moderately altered by collisional and non-gravitational effects. However, as I show in Chapter 2, the asteroids with $D \gtrsim 10\text{--}30$ km are also subject to weak chaotic evolution and escape from the main belt on gigayear timescales. By means of numerical simulations, the loss history of large asteroids in the main belt has been computed, as described in Chapter 2. I also computed the cumulative impacts of large asteroids on the terrestrial planets over the last ~ 3 Gy.

The orbital distribution of large asteroids that exist today must have been determined by dynamical processes in the early solar system, because large asteroids do not uniformly fill regions of the main belt that are stable over the age of the solar system (Minton and Malhotra, 2009, see also Chapter 3). The present study is therefore additionally motivated by the need to understand better the origin of the present dynamical structure of the main asteroid belt. The Jupiter-facing boundaries of some Kirkwood gaps are more depleted than the sunward boundaries, and the inner asteroid belt is also more depleted than a model asteroid belt in which only gravitational perturbations arising from the planets in their current orbits have sculpted an initially uniform distribution of asteroids. In Chapter 3 I show that the pattern of depletion observed in the main asteroid belt is consistent with the effects of resonance sweeping due to giant planet migration that is thought to have occurred early in solar system history (Fernandez and Ip, 1984; Malhotra, 1993, 1995; Hahn and Malhotra, 1999; Gomes et al., 2005), and that this event was the last major dynamical depletion event experienced by the main belt. The last major dynamical depletion event in the main asteroid belt likely coincided with the so-called Late Heavy Bombardment (LHB) ~ 3.9 Gy ago as indicated by the crater record of the inner planets and the Moon (**Strom et al., 2005**).

1.2   Planet migration and the Late Heavy Bombardment

Early numerical simulations of the formation of the outer ice giants, Uranus and Neptune, produced an unexpected result. In the simulations of **Fernandez and Ip (1984)**, the ice giants were grown from embryos 20% their present mass by accretion within a massive planetesimal disk, with Jupiter and Saturn at their present masses. They discovered that gravitational interactions between the gas giants, the ice giant embryos, and the planetesimal disk caused the orbits of the gas giants and ice giant embryos to migrate, with Jupiter migrating inward toward the Sun, and the outermost four large bodies migrating outward. This effect is called planetesimal-driven giant planet migration, and it can occur even if Uranus and Neptune have their present mass, but only if there is a massive planetesimal disk beyond the orbits of the giant planets. Planetesimal-driven planet migration can be understood in the following way.

A close encounter between a planetesimal and a giant planet can either increase or decrease the orbital angular momentum of the planetesimal and inversely that of the planet. Whether the planetesimal experiences an increase or a decrease in orbital angular momentum depends on details of the encounter, such as impact parameter and encounter angle. Due to the relatively small sizes of the ice giants, a planetesimal that experiences a close encounter with an ice giant is more likely to remain gravitationally bound to the Sun than escape. Because the planetesimal has experienced a close encounter with a planet, its orbit is strongly coupled to that planet and it will likely experience future close encounters with it. If a single ice giant were the only planet in the solar system, this would mean that any given planetesimal would encounter it over and over again, until chance would allow for an encounter that sent the planetesimal out of the solar system completely. If this process were repeated by large numbers of planetesimals, this would result in a net decrease in the ice giant's angular momentum, and its orbit would migrate sunward. However, the real solar system contains two ice giants, Uranus and Neptune, and two gas giants, Jupiter and Saturn. Planetesimals which have their angular momentum reduced by a close encounter with Neptune can potentially begin to encounter Uranus. Such a close encounter with Uranus can decouple the planetesimal from Neptune, reducing the chances for further close encounters with the outermost ice giant. In this way Uranus acts as a "sink" of planetesimals, and Neptune experiences a net increase in angular momentum and its orbit grows as it passes planetesimals inward toward Uranus.

A similar process occurs at Uranus, but in this case both Neptune and Saturn acts as planetesimal sinks. Because Saturn is larger than Neptune, Saturn is a more effective sink, and Uranus experiences a net increase in angular momentum after encounters with numerous planetesimals. And again at Saturn, Jupiter acts as a more effective sink than Uranus, so Saturn also migrates outwards. Therefore as the icy planetesimals interact with the giant planets, they tend to be passed down from one giant planet to its inward neighbor until they ultimately begin encountering Jupiter. There are no more giant planets inward of Jupiter to act as effective sinks. Also, because of Jupiter's large size, a planetesimal is much more likely to escape from the Sun on a hyperbolic trajectory after an encounter with Jupiter than with any of the other three giant planets. Jupiter therefore experiences a net decrease in angular momentum as it tosses planetesimals onto hyperbolic orbits sending them forever out of the solar system.

Evidence in the structure of the Kuiper Belt suggests that the early solar system experienced just such a phase of planetesimal-driven migration (Fernandez and Ip, 1984; Malhotra, 1993, 1995; Hahn and Malhotra, 1999; Levison et al., 2008). In particular, both the large eccentricity of Pluto and its location within the 3:2 mean motion resonance with Neptune are well explained by resonance capture during the outward migration of Neptune (Malhotra, 1993). The existence of populations of

Kuiper Belt Objects in the 3:2, 2:1, and other low order mean motion resonances with Neptune were predicted as an outcome of planetesimal-driven planet migration before these populations of resonant objects were discovered observationally (Malhotra et al., 2000).

A natural outcome of planetesimal-driven planet migration is to enhance the impact flux everywhere in the solar system. In the terrestrial planet region, planet migration can enhance the impact flux in two ways. First, the scattering of icy planetesimals by the giant planets would have resulted in many of those planetesimals crossing the orbits of the terrestrial planets. Second, because the dynamical structure of the main asteroid belt is dominated by the influence of the giant planets (most importantly Jupiter and Saturn), any change in the orbital properties of these planets in the past should have gravitationally disturbed the asteroid belt. As the giant planets migrated, locations of mean motion resonances as well as secular resonances would have swept across the asteroid belt, raising the eccentricities of asteroids to planet-crossing values. This period of planet migration has been suggested as a cause of the so-called Late Heavy Bombardment (LHB), assuming the onset of migration was sufficiently delayed (Gomes et al., 2005).

The Late Heavy Bombardment has long remained a puzzling and controversial topic in solar system chronology. The radiometric ages of ancient highland representing impact melt rocks returned during the American *Apollo* and Soviet *Luna* missions showed an apparent clustering at ~ 3.9 Gy ago (Turner et al., 1973; Tera et al., 1973, 1974; Ryder, 1990). One interpretation of the clustering of impact melt rock ages was that the collected samples represented the tail end of a long period of intense impact bombardment that began with the initial accretion of solids in the inner solar system 4.57 Gy ago (Hartmann, 1975; Russell et al., 2006). Hartmann (1975) suggested that the lack of ancient rocks (those with impact resetting ages $\gtrsim 4$ Gy) on the Moon is the result of a "stone-wall" effect, that is that prior to 4 Gy ago the impact rate was so high that all older impact melt rocks were obliterated into micron-size dust by subsequent impacts. The samples returned in the lunar missions of the late 1960s and early 1970s helped to calibrate crater counting-based age estimates of lunar surface features, which suggest that lunar cratering rapidly declined from a rate several orders of magnitude above the present value down to within a factor of two or three above the present value at $\sim$ 3.8–3.9 Gy ago (Ryder, 1990; Stöffler and Ryder, 2001; Hartmann et al., 2000; Ryder, 2002; Neukum et al., 2001). Under the stone-wall hypothesis, all ancient impact melt rocks have been destroyed beyond any ability to date, and the only remaining fragments are those produced at the end of this period of intense bombardment, $\sim$ 3.9 Gy ago. The crater densities of several ancient lunar terrains along with estimates of their absolute ages determined by radioisotope dating of samples returned by the *Apollo* and *Luna* missions is shown in Fig. 1.1.

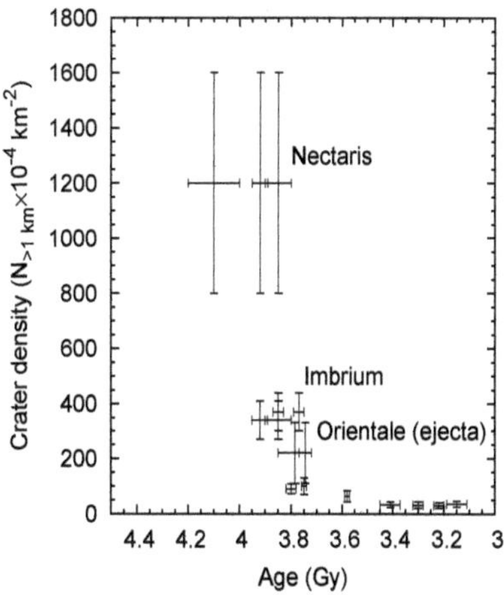

Figure 1.1: Crater densities vs. age for select lunar terrains. A rapid decline in the cratering rate at $\sim$ 3.9 Gy is observed. The age of Nectaris basin is controversial, and several proposed ages are shown here. Data taken from **Stöffler and Ryder** (2001).

Based on the resetting ages of several isotopic systems (U-Pb, K-Ar, and Rb-Sr) from lunar rock samples of ancient heavily cratered terrains, an alternative hypothesis, dubbed the "terminal lunar cataclysm," was suggested by members of the Caltech Lunatic Asylum. In this model, the Moon experienced a sudden and intense spike in its impact rate at $\sim$ 3.9 Gy ago (**Tera et al.**, 1974, 1973). Under this hypothesis, the lack of impact melts older than ~ 4.0 Gy is taken as evidence that the impact rate was relatively low in the interval between planetary accretion and 4 Gy ago (Chapman et al., 2007). In particular, unique characteristics of the U-Pb isotopic system suggest a cataclysmic event at ~ 4 Gy ago from the lunar samples, rather than simply the last stages of a monotonically declining impact rate stretching back to the beginning of the solar system. The lead isotope ^{207}Pb is a daughter product of the decay of ^{235}U, and ^{206}Pb is a product of the decay of ^{238}U, with different half lives for each. When a mineral assemblage becomes closed (that is, whatever lead or uranium exists in the rock or is subsequently produced by radioactive decay remains in the rock until it is analyzed), then a plot of ^{207}Pb$/^{206}$Pb vs. ^{238}U$/^{206}$Pb falls on a curve called the concordia, and the position on the curve is a function of the closure age (see Fig. 1.2). If at any point in the rock's history the system is reopened (that is, either uranium or lead is allowed to escape or accumulate within the rock) and then closed again, the resulting values of ^{207}Pb$/^{206}$Pb vs. ^{238}U$/^{206}$Pb will fall along a discordant line that intersects the concordia at the initial closure age and again at the age corresponding to the isotope mobilization event. Lead is a more volatile element than uranium, and therefore loss or accumulation of lead during heating of minerals drives the minerals to the discordant line. By measuring a variety of minerals that had varying amounts of either loss or accumulation of lead during a single heating event, the discordant line can be constructed, as shown in Fig. 1.2.

Tera et al. (1974) showed that a variety of *Apollo* samples show discordant U-Pb ratios that intersect the concordia at both ~ 4 Gy and ~ 4.45 Gy ago. Their interpretation for this result is that the lunar rocks became isotopically closed to U and Pb after the Moon formed around ~ 4.45 Gy ago. Then, a major global thermal event occurred at ~ 4 Gy ago that then opened the rocks to lead, allowing both radiogenic and non-radiogenic lead to escape some minerals and accumulate in others. The rocks have then remained closed to lead since this event. This scenario formed the basis of their lunar cataclysm argument, in which the impact rate on the Moon was low after its formation but then suddenly underwent a spike at ~ 4 Gy.

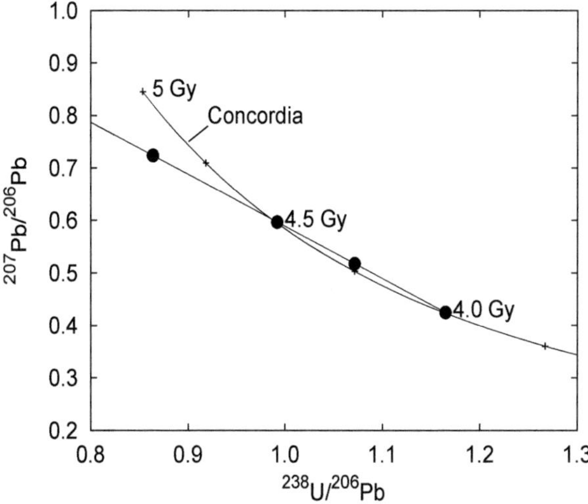

Figure 1.2: Illustration of a discordant U-Pb system. The points show measurements of a hypothetical assemblage of minerals in a rock that were closed to uranium and lead at 4.5 Gy ago and then reopened at 4 Gy ago. Radiogenic lead was depleted from some minerals and added to others, and the resulting data points plot along a discordant line that intersects the concordia at 4.5 Gy and 4 Gy. Compare with Fig. 3 of **Tera et al.** (1974) which shows data from Apollo samples.

However, the *Apollo* and *Luna* samples were obtained over a relatively small area of the lunar near side, and **Haskin et al.** (1998) has argued that the apparent clustering of ages seen in these samples results from ejecta from a single basin impact event, namely the formation of the Imbrium basin. Imbrium is the largest near side lunar basin. In contrast, **Norman et al.** (2006) have identified at least four separate heating events that cluster around 3.75–3.96 Gy ago on the basis of ^{40}Ar–^{39}Ar dating of a variety of *Apollo 16* impact melt breccias. They argue that each event seems to have unique petrological characteristics that correlate with each other in time, which implies that the clustering of ages is not the result of a single basin impact, but several over a relatively short period of time.

One way to determine whether the apparent cluster of ages is simply an artifact of the limited sampling area of lunar samples is analyze lunar meteorites. Lunar meteorites presumably originate from nearly anywhere on the Moon, and impact melts from several lunar meteorites show a lack of ages older than 3.92 Gy (**Cohen et al.**, **2000**, **2005**). However, rather than clustering at 3.9 Gy, the lunar meteorites impact melts show ages as young as 2.5 Gy. The meteorite record may also provide some insights into the bombardment rate of within the main asteroid belt. The class of meteorites called the H-chondrites record impact events in the ^{40}Ar-^{39}Ar system suggesting impact events within the first $\sim$ 100 My of the solar system (the era of planetary accretion), 3.6–4.1 Gy ago (the end of the LHB), but nothing in-between **Swindle et al.** (**2009**).

The hypothesis that the lunar cratering record represents a cataclysmic spike in the cratering rate on all the terrestrial planets (or perhaps even the entire solar system) of $\lesssim$ 100 My in duration, rather the tail end of the planetary accretion era that began with the condensation of the first solids 4.57 Gy ago, has remained a very controversial idea (**Hartmann et al.**, **2000**; **Chapman et al.**, **2007**). The differing hypotheses have very different implications for the early history of Earth. The Earth would have received an even higher rate of impacts than the Moon, due to its larger geometric and gravitational cross-sections. The end of the LHB at $\sim$ 3.8 Gy corresponds to the end of Earth's Hadean eon. The Hadean is the name given to the period of time on Earth between its formation at $\sim$ 4.5 Gy ago and the oldest known rocks at $\sim$ 3.9 Gy ago. The Hadean was so named because it was assumed that the Earth must have been a hellish place under such intense bombardment, where scarcely would the ocean begin to condense when a giant impact would vaporize it back into the atmosphere again (**Sleep et al.**, **1989**; **Chyba**, **1990**; **Nisbet and Sleep**, **2001**). Recently, however, evidence has emerged to contradict this picture. Small numbers of zircon crystals have recently been discovered that have U-Pb ages well within the Hadean. Zircon is an unusually hard mineral and can survive the intense heat and pressure that rocks experience as they undergo processing due to Earth's plate tectonics. Several of these zircon crystals suggest that continental crust and oceans were extent as early as 4.3 Gy ago (**Harrison**, **2005**; **Mojzsis et al.**, **2001**).

Under the declining impact hypothesis, Earth should not have had continental crust nor liquid water oceans at this time, as the impact rate would have been far too high (e.g., Ryder, 2002). In addition, some Hadean zircons which have formation ages older than 4 Gy ago show metamorphic overgrowths that date to 3.9 Gy (Trail et al., 2007). These overgrowths have been interpreted as representing short-lived heating events that reset the U-Pb system in the rim of the crystal, leaving the cores intact. These metamorphic overgrowths with ages of 3.94–3.97 Gy were seen in at least three separate crystals, however all crystals were collected at the same site and may represent a local event rather than a global one.

The hypothesis that LHB was simply the tail end of the initial accretion of the terrestrial planets is also problematic in light of post-*Apollo* era studies of terrestrial planet formation. Numerical modeling of formation of the terrestrial planets and the isotopic composition of the bulk Earth indicates that the accretion of the terrestrial planets happened on a very short timescale—on the order of a few tens of millions of years (Greenberg et al., 1978; Halliday et al., 2003; Kenyon and Bromley, 2006), and that the remnants of accretion should have been removed very quickly from the inner Solar System (in tens of millions of years rather than hundreds of millions) due to dynamics and collisional evolution (Bottke et al., 2007a). Thus it is very difficult to connect the intense bombardment observed in the rock record of the Moon some 700 My after the formation of the planets with planetary accretion itself. Both the evidence of the existence of Hadean continents and oceans, and the short timescale for the depletion of accretion remnants from the inner Solar System, imply that the Earth's bombardment rate was likely much lower during most of the Hadean than is implied by a steadily decaying bombardment rate. Therefore the LHB may indeed have been a cataclysmic event that took place over a relatively short duration ending at about 3.8–4.0 Gy ago.

Strom et al. (2005) showed that the heavily cratered terrains on the Moon, Mars, and Mercury that are presumed to date to the time of the LHB all have size-frequency distributions that are consistent with impactors originating in the main asteroid belt. This evidence suggests that, at least by the end of the LHB, the impactors that dominated the cratering record on the terrestrial planets were ejected into terrestrial planet-crossing orbits from the main asteroid belt by a size-independent mechanism. Resonance sweeping by planetesimal-driven giant planet migration is a compelling mechanism for exciting main belt asteroids into planet-crossing orbits in a size-independent way. A problem with the hypothesis that giant planet migration caused the LHB is that the timing of the LHB puts the event 600–800 My after the formation of the solar system. The timing and duration of planet migration is uncertain, but one successful model of planet migration, the so-called "Nice model" has, through a suitable choice of initial conditions of the solar system, demonstrated the ability to delay the onset of migration until up to ~ 1 Gy after the formation of the solar system (Gomes et al., 2005). The Nice model therefore provides a plausible mechanism for delaying giant planet migration such that it coincides with the LHB, but the exact timing of the destabilizing trigger at ~ 700 My after planet formation is not a necessary outcome. As I will show in Chapter 5, planet migration itself likely took place over a very time span ($\lesssim 10^7$ yr), and therefore any model which proposes to link planetesimal-driven planet migration with the LHB must include a mechanism to delay the onset of migration for several 10^8 yr. Regardless of the controversies surrounding the cause and duration of the LHB, the hypothesis that the giant planets experienced a phase of planetesimal-driven migration is well supported by observations, and in Chapter 3 I show that patterns of depletion observed in the asteroid belt are consistent with the effects of sweeping of resonances during the migration of the outer giant planets.

1.3   Overview of the present work

Motivated by the compelling link between planetesimal-driven planet migration, the dynamical history of the asteroid belt, and the LHB, I seek to quantify the dynamical effects that planet migration produced on the early asteroid belt. I begin by looking for clues in the present-day structure of the observed asteroid belt, using observations to infer the post-LHB dynamical history of asteroids. I describe in detail the dynamical models of the present-day asteroid belt in Chapter 2. These models are constrained by the observed orbital distributions of large asteroids in the main belt. I use these models to explore the dynamical mechanism by which large asteroids have been lost from the main belt over the last 4 Gy, and show that the impact rate of large ($D > 10$ km) asteroids onto the terrestrial planets may be significantly underestimated. In Chapter **3** I use the results of the asteroid belt models developed in Chapter 2 to show that asteroids currently do not uniformly fill all of the stable regions of the asteroid belt. Much of Chapter **3** has been published as **Minton and Malhotra (2009)**.

Asteroid eccentricity excitation by the sweeping of the ν_6 resonance is the primary mechanism by which most asteroids were removed from the asteroid belt and placed onto planet-crossing orbits during the epoch of planetesimal-driven planet migration. In order to better understand the process by which asteroid depletion occurred during planet migration, I have developed an analytical model of the sweeping ν_6 secular resonance. In Chapter 4 I show how the secular dynamics of the solar system is changed when Jupiter and Saturn are displaced from their current semi-major axes. I use the method of Fourier analysis of long-term integrations of the two giant planets to show how the magnitude of the g_6 eigenfrequency (which determines the position of the ν_6 resonance) changed as a function of the giant planets' migration history. I also show that a relatively simple secular theory is an adequate model of the change in the g_6 eigenfrequency as a function of Saturn's position for the majority of Jupiter and Saturn's migration history. The secular theory is based on the Laplace-Lagrange linear secular theory with the correction due to the proximity of Jupiter and Saturn to their 2:1 mean motion resonance that was developed by **Malhotra et al. (1989)**. I use results derived in Chapter 4 as inputs to a model of the excitation of asteroids by the sweeping ν_6 resonance, in Chapter 5. The analytical model of the sweeping ν_6 resonance explicitly relates an asteroids change in eccentricity excitation to its initial eccentricity and the migration rate of Saturn. I use the results of the analytical model to set an upper limit on the rate of migration of Saturn. I also show that a peculiar feature of the eccentricity distribution of asteroids in the main belt is consistent with the effects of the sweeping ν_6, and may help further constrain the migration rate of Saturn.

Establishing a link between the epoch of planet migration and the Late Heavy Bombardment may require further observational tests. In Chapter 6 I explore whether "pollution" of the solar atmosphere by lithium may have left its trace in the lunar rock record. Lithium is an element that is more rare in the Sun than in planetesimals because lithium is destroyed by thermonuclear fusion at relatively low temperatures. I model the abundance of solar lithium and the ratio of two isotopes of lithium, ^{6}Li and ^{7}Li, under a variety of assumptions. These models may also help observationally identify exosolar systems experiencing their own versions of the LHB.

I include here a table of all symbols used and their basic definitions.

Table 1.1: Symbols and their definitions

Symbol	Definition
	Dynamics and orbital elements
G	Universal gravitational constant
a	Semimajor axis
e	Eccentricity
i	Inclination
ϖ	Longitude of pericenter
Ω	Longitude of the ascending node
λ	Mean longitude
J	Conjugate momentum $\sqrt{a}(1 - \sqrt{1 - e^2})$
τ	e-folding timescale
g_i	Secular eigenfrequency of planet i
β_i	Phase of the eigenmode of planet i
$E_j^{(i)}$	Amplitude of the jth eigenmode in planet i
	Projectile and crater physical properties
D	Diameter (of projectile in the context of cratering)
D_c	Final crater diameter
H	Absolute visual magnitude
m	Mass
ρ	Density
ρ_v	Geometric visual albedo
θ	Impact angle
N	Number

CHAPTER 2

DYNAMICAL EROSION OF THE ASTEROID BELT

2.1 Introduction

Knowledge about the distribution of the asteroids in the main belt just after that last major depletion event may help constrain models of that event. Quantifying the dynamical loss rates from the asteroid belt also help us understand the history of large impacts on the terrestrial planets. Motivated by these considerations, in this chapter I explore the dynamical erosion of the main asteroid belt, which is the dominant mechanism by which large asteroids have been lost over the last ~ 4 Gy.

I have performed a series of n-body simulations of large numbers of test particles in the main belt region over long periods of time (4 Gy and 1.1 Gy). I derive an empirical functional form for the population decay and loss rate of main belt asteroids. Finally, I discuss the implications of my results for the history of large asteroidal impacts on the terrestrial planets.

2.2 Numerical simulations

My long term orbit integrations of the solar system used a parallelized implementation of a second-order mixed variable symplectic mapping known as the Wisdom-Holman Method (**Wisdom and Holman, 1991**; **Saha and Tremaine, 1992**), where only the massless test particles are parallelized and the massive planets are integrated in every computing node. My model included the Sun and the planets Mars, Jupiter, Saturn, Uranus, and Neptune. All masses and initial conditions were taken from the JPL Horizons service[1] on July 21, 2008. The masses of Mercury, Venus, Earth, and the Moon were added to the mass of the Sun.

---

[1]see http://ssd.jpl.nasa.gov/?horizons

Test particle asteroids were given eccentricity and inclination distributions similar to the observed main belt, but a uniform distribution in semimajor axis. The initial eccentricity distribution of the test particles was modeled as a Gaussian with the peak at $\mu = 0.15$ and a standard deviation of 0.07, a lower cutoff at zero, and an upper cutoff above any value which would lead to either a Mars or Jupiter-crossing orbit, whichever was smaller. The initial inclination distribution was modeled as a Gaussian with the peak $\mu = 8.5°$ and standard deviation of $7°$, and a lower cutoff at $0°$. The other initial orbital elements (longitude of ascending node, longitude of perihelion, and mean anomaly) were uniformly distributed. The eccentricity and inclination distributions of the adopted initial conditions and those of the observed asteroids of absolute magnitude $H \leq 10.8$ are shown in Fig. 2.1.

Mars was the only terrestrial planet integrated in my simulations. Despite its small mass, Mars has a significant effect on the dynamics of the inner asteroid belt due to numerous weak resonances, including three-body Jupiter-Mars-asteroid resonances (Morbidelli and Nesvorny, 1999). Two simulations were performed: Sim 1, with 5760 test particles integrated for 4 Gy, and Sim 2, with 115200 test particles integrated for 1.1 Gy. In each of these simulations an integration step size of 0.1 yr was used. Particles were considered lost if they approached within a Hill radius of a planet, or if they crossed either an inner boundary at 1 AU or an outer boundary at 100 AU.

I define time $t = 0$ as the epoch when the current dynamical architecture of the main asteroid belt and the major planets was established. What I mean by this is the time at which any primordial mass depletion and excitation has already taken place (see O'Brien et al., 2007), and any early orbital migration of giant planets has finished (Fernandez and Ip, 1984; Malhotra, 1993; Strom et al., 2005; Gomes et al., 2005; Minton and Malhotra, 2009). At this epoch the main belt would have already had its eccentricity and inclination distributions excited by some primordial process, and its semimajor axis distribution shaped by early planet migration. Therefore, the e and i distributions at $t = 0$ likely resembled those of the present-day asteroid belt, although subsequent long-term evolution likely altered them somewhat from

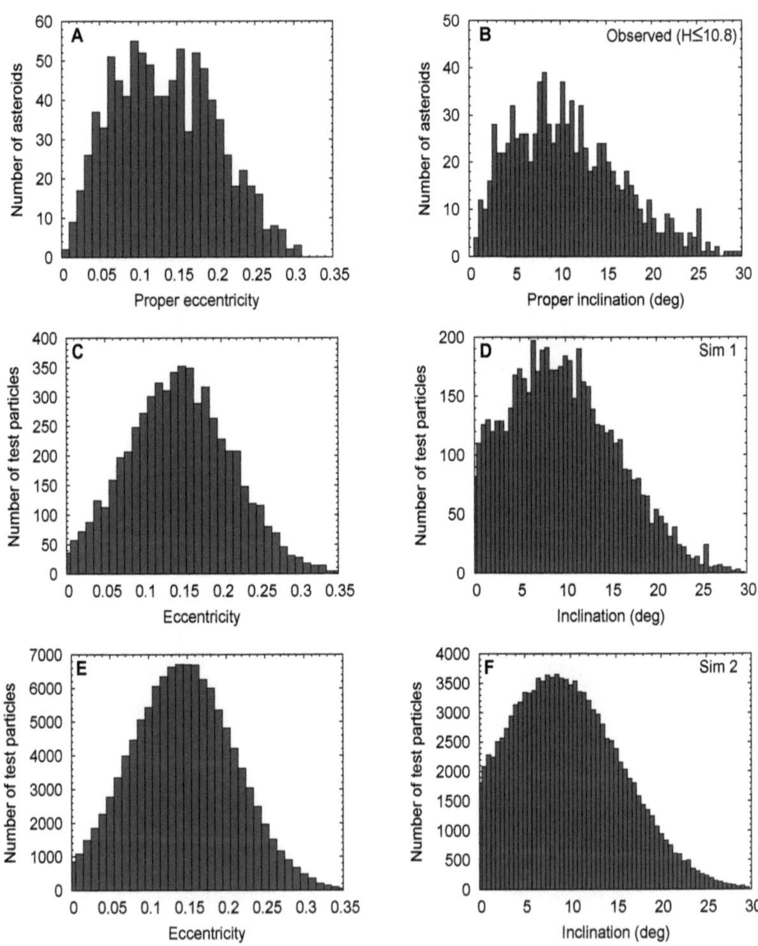

Figure 2.1: Eccentricity and inclination distributions. **a),b)** The distributions of the 931 observed bright ($H \leq 10.8$) asteroids in the main belt that are not members of collisional families, from the AstDys online data service (**Knežević and Milani, 2003; Nesvorný et al., 2006**). **c),d)** The initial e and i distributions for Sim 1 (the 5760 particle simulation). **e),f)** The initial e and i distributions for Sim 2 (the 115200 particle simulation).

their primordial state. Current understanding of planetary system formation suggests that the epoch prior to when I define $t = 0$ could have been several million to several hundred million years subsequent to the formation of the first solids in the protoplanetary disk; the first solids have radiometrically determined ages of 4.567 Gy (Russell et al., 2006).

2.3   Main asteroid belt population evolution

The loss history of particles from Sim 1 and Sim 2 are shown in Fig. 2.2. The loss histories are nearly indistinguishable over the 1.1 Gy length of Sim 2. The loss history appears to go through two phases. The first phase, lasting until ~ 1 My, is characterized by a rapid loss of particles from highly unstable regions, such as the major Kirkwood gaps and the ν_6 secular resonance. The slope of the loss rate on a log-linear scale changes rapidly between 0.3–1 My until the second phase is reached, which lasts from 1 My until at least the end of Sim 1 at 4 Gy. The slope of the loss rate on a log-linear scale continues to change during the second phase, but only over much longer timescales and by a much smaller amount than during the first phase.

The particle removal times and particle fates (whether they become inward-going Mars-crossers, or outward-going Jupiter-crossers) for Sim 2 are both shown in Fig. 2.3. I find that the particles that are lost during the initial 1 My (the red to light-green points) are generally those with high initial eccentricity, particles from the ν_6 resonance (appearing as a curving yellow band in the semimajor axis vs. inclination plot), and particles in the strongly chaotic mean motion resonances with Jupiter (the Kirkwood gaps). These maps are similar to those produced by Michtchenko et al. (2009), however their maps were coded by spectral number (more chaotic orbits having a larger spectral number) using 4.2 My integrations. The apparent rapid change in slope around 10^5–10^6 years is likely due mostly to the emptying of asteroids from the ν_6 resonance region (see also the upper right-hand panel of Fig. 2.3). Also, in the outer asteroid belt the sheer proximity to Jupiter and the resulting strong short-term perturbations cause particles to be lost very rapidly.

Many of these regions may never have accumulated asteroids, and therefore the loss of these particles represents a numerical artifact in the simulation due to over-filling the model asteroid belt with test particles. For instance, if the asteroid belt formed with the giant planets in their current positions, those regions would always have been unstable to asteroids, so none could have formed there. However, regions of the asteroid belt that are currently highly unstable may not always have been. Models of early solar system history indicate large changes to the orbital properties of the giant planets (Fernandez and Ip, 1984; Hahn and Malhotra, 1999; Tsiganis et al., 2005). This "numerical artifact" is useful in indicating that the timescales of clearing in the strongly unstable zones is $\lesssim 1$ My. Below I discuss in detail the loss of asteroids from the more stable regions of the main belt.

2.3.1 Historical population of large asteroids.

I used the test particle loss history of Sim 1 to estimate the loss history of large asteroids from the main belt and the large asteroid impact rate on the terrestrial planets. To do this I scaled f, the fraction of surviving particles in Sim 1 at $t = 4$ Gy, to the number of large asteroids in the current main belt. For this purpose, I define "large asteroid" as an asteroid with $D > 30$ km. For most asteroids, size is not as well determined as absolute magnitude. If the asteroid visual albedo, ρ_v, is known, the absolute magnitude can be converted into a diameter with the following formula (Fowler and Chillemi, 1992):

$$D = \frac{1329 \text{ km}}{\sqrt{\rho_v}} 10^{-H/5}.$$ (2.1)

Because asteroids can have a range of albedos, converting from brightness to diameter is fraught with uncertainty in the absence of albedo measurements. For simplicity, I adopt a single albedo, $\rho_v = 0.09$, which is approximately representative over the size range of objects considered here (Bottke et al., 2005a). In subsequent analysis I will use absolute magnitude as a proxy for size.

Using Eq. (2.1) and the assumption of albedo, an asteroid of diameter $D = 30$ km has an absolute magnitude $H = 10.8$. The main belt is observationally complete

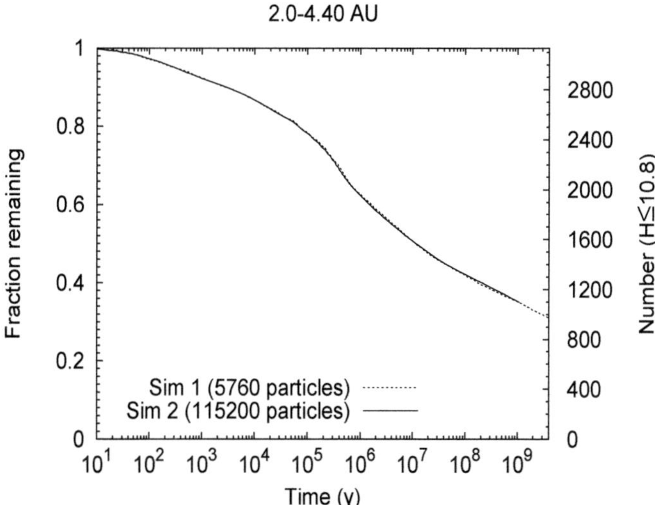

Figure 2.2: Loss history of test particles in the main asteroid belt region of the solar system from both Sim 1 (5760 particles for 4 Gy) and Sim 2 (115200 particles for 1.1 Gy). The left-hand axis is the fraction of the original test particle population that have survived the simulation at a given time. The right-hand axis is the estimated number of large asteroids in the asteroid belt, and is computed by equating the fraction remaining at $t = 4$ Gy with the number of observed $H \leq 10.8$ asteroids. The observational sample used is the 931 asteroids with $H \leq 10.8$ excluding members of collisional families.

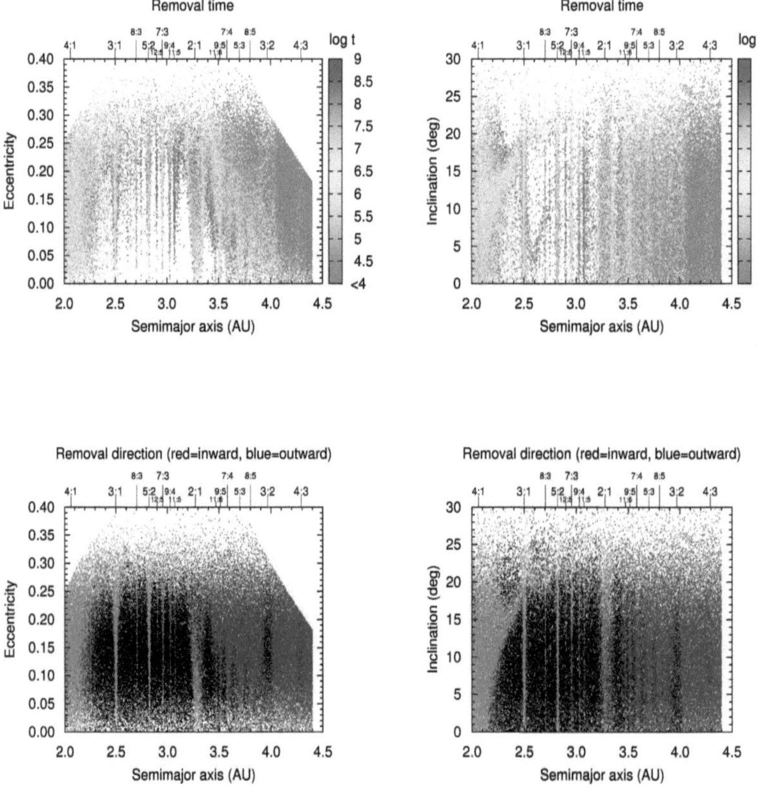

Figure 2.3: Removal statistics of the test particles in Sim 2 (115200 particles) as function of their initial orbital elements. In the upper panels, the points are colored to indicate their lifetime in the simulation, with red being the shortest-lived particles and blue being the longest-lived particles (particles surviving at the end of the simulation were removed for clarity). In the lower panels, the points are colored to indicate the direction in which they are lost: red indicates loss due to either a close encounter with Mars or removal at the inner boundary at 1 AU, blue indicates loss due either a close encounter with a giant planet or removal at the outer boundary at 100 AU, and black indicates particles that survived the entire 1.1 Gy simulation.

for asteroid absolute magnitudes as faint as $H = 13$ (Jedicke et al., 2002). While most $H \leq 10.8$ asteroids have existed relatively unchanged over the last 4 Gy, a few breakup events have created some large fragments over this timespan. For example, there are five members of the Vesta family with $H < 10.8$ (Nesvorný et al., 2006). Collisional fragments produced over the last 4 Gy can "contaminate" the observed $H \leq 10.8$ asteroid population, and collisional breakup events have also disrupted some primordial $H \leq 10.8$ asteroids. These collisional processes complicate the estimate of the dynamical loss history of large asteroids over the age of the solar system. Happily, most collisional family members with $H \leq 10.8$ have been identified (Nesvorný et al., 2006), and can be removed to further refine the observation sample. The database of (Nesvorný et al., 2006) was also made with some attempt to remove interlopers that have similar dynamical properties as a family, but a different spectral classification that indicates they are not members of the collisional family (Mothé-Diniz et al., 2005)

The observational data set I used was the 1137 asteroids with $H \leq 10.8$ obtained from the AstDys online data service (Knežević and Milani, 2003). Using the family classification system of Nesvorný et al. (2006), 206 (18%) of these asteroids are identified members of collisional families. I eliminated collisional family members and used the remaining sample of 931 asteroids for the scaling. Implementing this normalization, Fig. 2.2 shows the loss history of the main belt asteroids with the population scale on the right-hand axis. Because the dynamical depletion of asteroids from the main belt is approximately logarithmic, a roughly equal amount of depletion occurred in the time interval 10–200 My as in 0.2–4 Gy. I find that the asteroid belt at $t = 200$ My would have had 28% more large asteroids than today, and the asteroid belt at $t = 10$ My would have had 64% more large asteroids than today. My calculation indicates that ~ 2200 large asteroids ($H \leq 10.8$) may have been lost from the main asteroid belt by dynamical erosion since the current dynamical structure was established, but ~ 1600 of those asteroids would have been lost within the first 10 My.

2.3.2 Non-uniform pattern of depletion of asteroids

Fig. 2.4 compares the results of Sim 1 with the observational sample ($H \leq 10.8$ asteroids, excluding collisional family members).[2] The proper semimajor axes of the surviving particles from Sim 1 at the end of the 4 Gy integration were computed using the public domain Orbit9[3] code (**Knezevic et al., 2002**). The bin size of 0.015 AU was chosen using the histogram bin size optimization method described by **Shimazaki and Shinomoto (2007)** (See A). Fig. 2.4b is the ratio of the data sets. I find that the observed asteroid belt is overall more depleted than the dynamical erosion of an initially uniform population can account for, and there is a particular pattern in the excess depletion: there is enhanced depletion just exterior to the major Kirkwood gaps associated with the 5:2, 7:3, and 2:1 mean motion resonances (MMRs) with Jupiter (the regions spanning 2.81–3.11 AU and 3.34–3.47 AU in Fig. 2.4a); the regions just interior to the 5:2 and the 2:1 resonances do not show significant depletion (the regions spanning 2.72–2.81 AU and 3.11–3.23 AU in Fig. 2.4a), but the inner belt region (spanning 2.21–2.72 AU) shows excess depletion.

Minton and Malhotra (2009) showed that the observed pattern of excess depletion is consistent with the effects of the sweeping of resonances during the migration of the outer giant planets, most importantly the migration of Jupiter and Saturn. There is evidence in the outer solar system that the giant planets – Jupiter, Saturn, Uranus and Neptune – did not form where we find them today. The orbit of Pluto and other Kuiper Belt Objects (KBOs) that are trapped in mean motion resonances with Neptune can be explained by the outward migration of Neptune due to interactions with a more massive primordial planetesimal disk in the outer regions of the solar system (**Malhotra, 1993, 1995**). The exchange of angular momentum between planetesimals and the four giant planets caused the orbital migration of the giant

---

[2]Fig. 2.4a is similar to Fig. 1a of **Minton and Malhotra (2009)**, but with my sample of 931 asteroids with $H \leq 10.8$ that are not members of collisional families. The results I report in this section are similar to those in Chapter **3**; because they are based on simulations with much larger number of particles, their statistical significance is improved.

[3]Found at: http://hamilton.dm.unipi.it/astdys/

planets until the outer planetesimal disk was depleted of most of its mass, leaving the giant planets in their present orbits (Fernandez and Ip, 1984; Hahn and Malhotra, 1999; Tsiganis et al., 2005). As Jupiter and Saturn migrated, the locations of mean motion and secular resonances swept across the asteroid belt, exciting asteroids into terrestrial planet-crossing orbits, thereby greatly depleting the asteroid belt population and perhaps also causing a late heavy bombardment in the inner solar system (Liou and Malhotra, 1997; Levison et al., 2001; Gomes et al., 2005; Strom et al., 2005).

I identified six zones of excess depletion; these are labeled I–VI in Fig. 2.4b. Zone I would have experienced depletion primarily due to the sweeping ν_6 resonance, with some contribution possibly from the 3:1 MMR. Zones II and IV are the zones that lie on the sunward sides of the 5:2 and 2:1 resonances, respectively, and are hypothesized to have experienced the least amount of depletion due to sweeping mean motion resonances (MMRs) and secular resonances under the interpretation of Minton and Malhotra (2009). Zones III, V, and VI are on the Jupiter-facing sides of the 5:2, 7:3, and 2:1 MMRs, respectively, and are hypothesized to have experienced depletion due to the sweeping of these resonances. The average ratio between the model and observed population per 0.015 AU bin in each zone is quantified in Fig. 2.5.

2.3.3 Empirical models of population decay

The loss rate of small bodies from various regions of the solar system has been studied by several authors (see Dobrovolskis et al., 2007, for a comprehensive review of the recent literature on the subject). Holman and Wisdom (1993) found that the decay of a population of numerically integrated test particles on initially circular, coplanar orbits distributed throughout the outer solar system was asymptotically logarithmic, that is, $\dot{n} \propto t^{-1}$, where n is the number of test particles remaining in the simulation at a given time t. Dobrovolskis et al. (2007) showed that, for many small body populations, loss is described as a stretched exponential decay, given by the Kohlrausch formula,

$$f = \exp\left(-[t/t_0]^\beta\right), \tag{2.2}$$

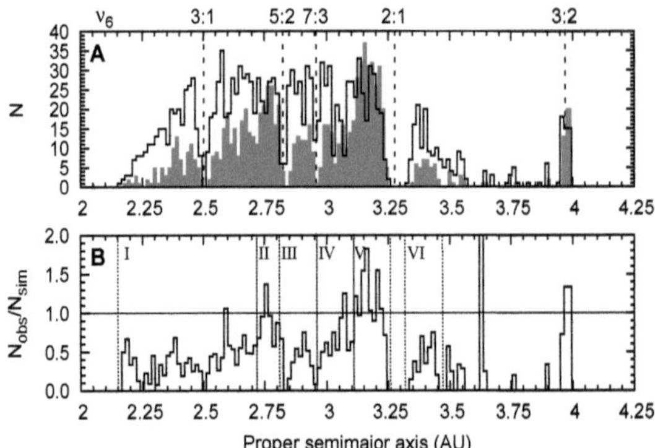

Figure 2.4: Comparison between the semimajor axis distribution of Sim 1 test particles and the sample of observed main belt asteroids. **a)** The observed main belt asteroid distribution, N_{obs}, for $H \leq 10.8$ asteroids (shaded) and the surviving particles of Sim 1, N_{sim}, (solid). **b)** Ratio of the data sets.

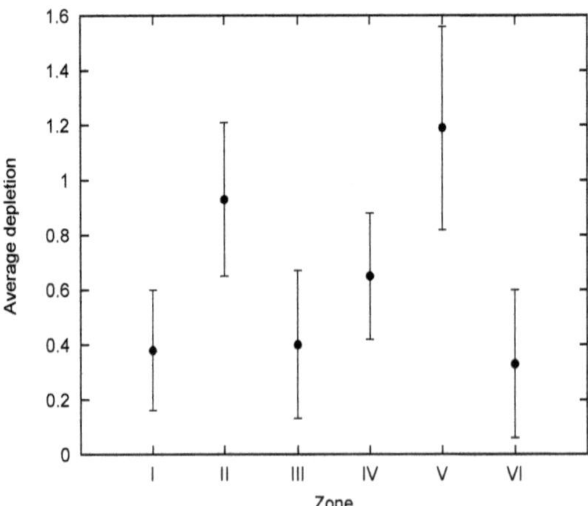

Figure 2.5: The average depletion, $\langle N_{obs}/N_{sim} \rangle$, in each of the six zones identified in Fig. 2.4b; the average is taken over the 0.015 AU bins present in each zone.

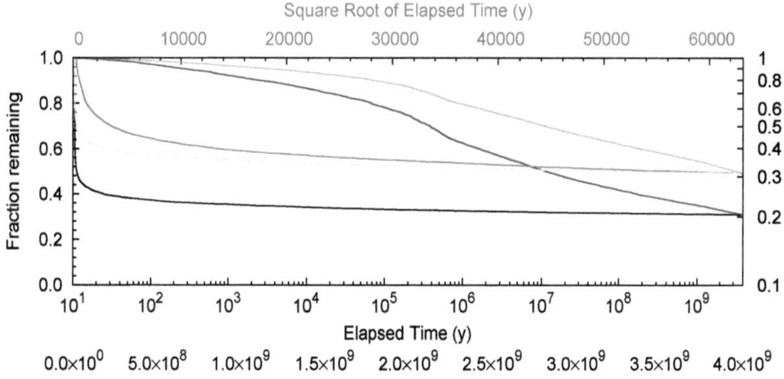

Figure 2.6: Comparison of empirical decay laws for the main asteroid belt region from Sim 1 (5760 particles for 4 Gy). The result from Sim 1 is plotted in five different ways. Bottom curve (black): fraction f of particles surviving (left-hand scale) vs. time (bottom scale). Next-to-lowest curve (green): $\log f$ (right-hand scale) vs. elapsed time (bottom scale). Middle curve (red): $\log f$ (right-hand scale) vs. $\sqrt{t}$ (top-scale). Next-to-uppermost curve (blue): f (left-hand scale) vs. $\log t$ (interior scale). Top curve (yellow): $\log f$ (right-hand scale) vs. $\log t$ (interior scale). Only the yellow and blue curves resemble straight lines in this format, and only for $t \gtrsim 10^6$ yr.

where f is the fraction remaining of the initial population ($f = n(t)/N_{tot}$). In the cases that **Dobrovolskis et al.** studied, namely loss rates of small body populations orbiting giant planets, they found that $\beta \approx 0.3$. For reference, note that a value of $\beta = 1/2$ is expected for a diffusion-dominated process for the removal of particles; in this case a plot of log of the number of remaining particles vs. square root time would be a straight line.

In Fig. 2.6 I adopt a similar plot style as in **Dobrovolskis et al.** (2007) (their Fig. 1) as a way of evaluating various empirical decay laws for the results of Sim 1. Unlike the cases explored by **Dobrovolskis et al.**, stretched exponential decay with $\beta \sim 0.3$–0.5 is a very poor model for the asteroid belt. Fig. 2.6 suggests either logarithmic or power law decay would be better models of particle decay from this simulation for $t > 10^6$ yr. Using a logarithmic decay law of the form:

$$f = A - B \ln(t/1 \text{ yr}), \tag{2.3}$$

and a power law decay of the form:

$$f = C(t/1 \text{ yr})^{-D}, \tag{2.4}$$

where A, B, C, and D are positive and dimensionless constants, I can look for the logarithmic, power law, and stretched exponential functions that best fit the decay history of Sim 1 for $t > 1$ My. The best fit parameters for each of these decay laws are listed in Table 2.1. Note that the best fit exponent D for the power law decay is close enough to zero that it is not very different than the logarithmic decay over the range of timescales considered here.

Table 2.1: Best fit decay laws for Sim 1 (5760 particles for 4 Gy)

Decay law	Parameters	Valid range (yr)
Stretched exponential (Eq. 2.2)	$\log t_0 = 8.6986 \pm 0.0057$	$t > 10^6$
	$\beta = 0.1075 \pm 0.0004$	
Logarithmic (Eq. 2.3)	$A = 1.1230 \pm 0.0020$	$t > 10^6$
	$B = 0.0377 \pm 0.0001$	
Power law (Eq. 2.4)	$C = 1.9556 \pm 0.0027$	$t > 10^6$
	$D = 0.0834 \pm 0.0001$	
Piecewise logarithmic (Eq. 2.5)	$A_1 = 1.3333 \pm 0.0006$	$10^{6.0} < t < 10^{7.2}$
	$B_1 = 0.05130 \pm 0.00004$	
	$A_2 = A_1 + (B_2 - B_1) \cdot 7.2$	$10^{7.2} < t < 10^{8.3}$
	$B_2 = 0.02695 \pm 0.00011$	
	$A_3 = A_2 + (B_3 - B_2) \cdot 8.3$	$10^{8.3} < t < 10^{9.1}$
	$B_3 = 0.02695 \pm 0.00011$	
	$A_4 = A_3 + (B_4 - B_3) \cdot 9.1$	$10^{9.1} < t < 10^{9.6}$
	$B_4 = 0.03079 \pm 0.00018$	

The difference between various empirical decay laws and the simulation output from Sim 1 is shown in Fig. 2.7. The format of Fig. 2.7 is similar to that of Fig. 2 of Dobrovolskis et al. (2007), but here the y-axis is $\Delta \log |\ln f| = \log |\ln f_{sim}| - \log |\ln f_{fit}|$, where the subscripts sim and fit refer to the simulation data and best fit model, respectively. A perfect fit would plot as a straight line with $\Delta \log |\ln f| = 0$. The power law function is a good fit, but with a small exponent D, which makes it practically similar to a logarithmic decay. The best fit stretched exponential value of β obtained here is much smaller than the value of ~ 0.3–0.5 obtained by many of the cases shown by Dobrovolskis et al. (2007). This may indicate that classical diffusion does not dominate the loss of asteroids from the main belt.

From Fig. 2.7 there is no clear preference for one or the other functions for the decay model. However, with some experimentation, I found that an improved fit can be obtained by considering a piecewise logarithmic decay of the form:

$$f_i = A_i - B_i \ln(t/1 \text{ yr}), \ t_i < t < t_{i+1}, \tag{2.5}$$

where A_i and B_i are positive coefficients. A physical justification for a piecewise logarithmic decay law is outlined in the following argument. If the region under study were divided into smaller subregions, and the loss of particles from each of those subregions follows a logarithmic decay law, then the linear combination of the decay laws for all subregions is the decay law for the total ensemble of particles, and is itself logarithmic. However, if any subregion completely empties of particles, then that region remains empty (it cannot have negative particles), and so the decay law of that region no longer contributes to the decay law for the total ensemble of particles; the decay of the ensemble then undergoes an abrupt change in slope. A piecewise logarithmic decay law for an ensemble of particles originating from the main asteroid belt region implies that the intrinsic loss rate from the asteroid belt is best described as $\dot{n} \propto t^{-1}$, but with different proportionality constants for different regions inside the belt.

I found that the model that minimized $\Delta \log |\ln f|$ for $t > 10^6$ yr is a four component piecewise logarithmic decay law with slope changes at $10^{7.4}$ yr, $10^{8.3}$ yr, and $10^{9.1}$ yr. I performed a least-squares fit to the loss history of Sim 1, fitting it to the four component piecewise function given by Eq. (2.5); the best fit parameters are given in Table 2.1. The residuals for the piecewise logarithmic decay law are much reduced, compared to the other empirical models considered, as shown in Fig. 2.7.

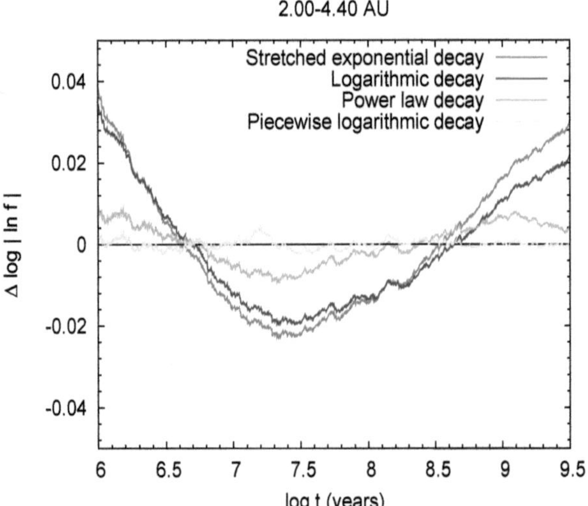

Figure 2.7: Differences between the best fit decay models and the loss history of Sim 1. Here the y-axis is $\Delta \log |\ln f| = \log |\ln f_{sim}| - \log |\ln f_{fit}|$, where the subscripts *sim* and *fit* refer to the simulation data and best fit loss function, respectively.

2.4   The effect of Mars

Morbidelli and Nesvorny (1999) showed that the planet Mars is responsible for contributing to orbital chaos of asteroids through weak mean motion resonances as well as three-body resonances. Sim 1b used the same inputs as Sim 1, but did not include the planet Mars directly, but simulated its effect indirectly by increasing the inner cutoff distance to 1.5 AU. This allows us to roughly distinguish the full long-range gravitational perturbing effects of Mars on the asteroid belt from the effects of only close encounters with that planet. Comparisons between loss histories between Sims 1 and 1b for the entire asteroid belt as well as 0.8 AU wide subdivisions are shown in Fig. 2.8. The best fit logarithmic law decay for Sim 1b has the slope $B = 0.0261$, compared with $D = 0.0377$ for Sim 1. The full effects of Mars increased the total number of particles lost from of the main belt region over the time interval 1 My–4 Gy by 8%. As Fig. 2.8 illustrates, the additional loss is primarily confined to the inner asteroid belt. This indicates that the effect of Mars due to distant gravitational perturbations is more potent to the loss of asteroids than its effect as a inner barrier alone.

2.5   Large asteroid impacts on the terrestrial planets

Although the impact history of the terrestrial planets is numerically dominated by small impactors, $D \lesssim 10$ km, the larger but infrequent impactors are also of great interest as they cause the more dramatic geological and environmental consequences. The dynamical origins of the latter have been less well studied because of the unavoidable small number statistics issues with them. Unlike the Yarkovsky effect, which is primarily responsible for populating the NEA population with $D \lesssim 10$ km asteroids, the dynamical chaos in the asteroid belt is a size independent process and dynamical erosion is the primary loss mechanism for large asteroids. Migliorini et al. (1998) investigated how $D > 5$ km asteroids from the main belt become terrestrial planet-crossing orbits, and found that weak resonance in the inner solar system were likely responsible for populating the NEOs. I used my simulations to quantify the

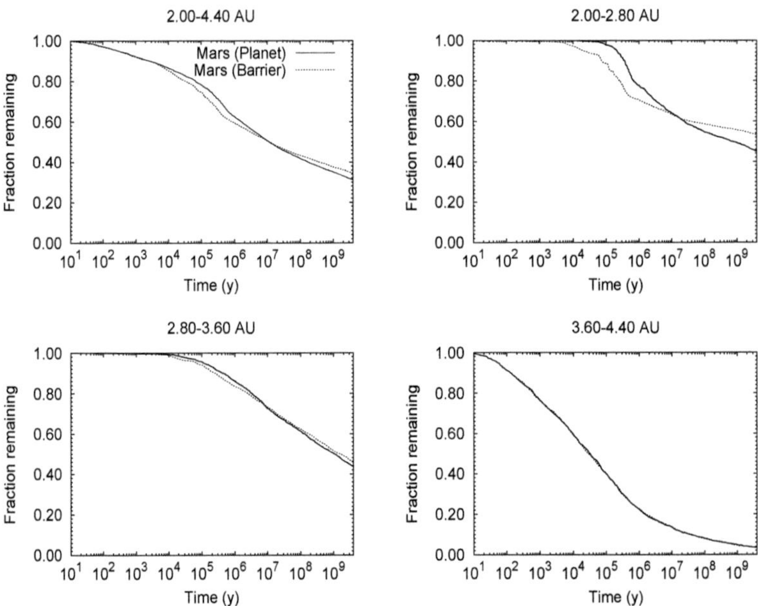

Figure 2.8: Loss history of test particles in the main asteroid belt region of the solar system from Sim 1 (5760 particles for 4 Gy) and for the variant Sim 1b (The same input as Sim 1 but without Mars and the inner cutoff distance moved up to 1.5 AU. The top left panel is for the entire ensemble of particles from 2.0–4.4 AU. The top right, bottom left, and bottom right panels are for inner, middle, and outer subregions 0.8 AU in width, respectively. The gravitational perturbations due to Mars cause the overall loss to be steeper than otherwise, with the most pronounced difference in the inner region of the main belt. The outer region of the main belt is not affected by perturbations from Mars.

impact rates for the larger asteroidal impactors.

As shown in §2.3.3, the long term dynamical loss of asteroids from the main belt is nearly logarithmic in time. The fate of any particular asteroid (the probability that it will impact a particular planet, the Sun, or be ejected from the solar system) is strongly dependent on its source region in the main belt (Morbidelli and Gladman, 1998; Bottke et al., 2000). For example, particles originating from the region near the ν_6 secular resonance at the inner edge of the main belt have a $\sim$ 1–3% chance of impacting the Earth (Morbidelli and Gladman, 1998; Ito and Malhotra, 2006), whereas objects originating further out in the asteroid belt have much lower Earth impact probabilities (Gladman et al., 1997). My simulations indicate that large asteroidal impactors that enter the inner solar system may originate throughout the main belt, so I needed to compute the overall impact probabilities for impactors originating by dynamical chaos from the main belt as a whole. I do this by means of an additional numerical simulation that yields the terrestrial planet impact statistics for those particles of Sim 1 that were 'lost' to the inner solar system. I then combine the impact probabilities with the 4 Gy loss history of large asteroids (Fig. 2.2) to estimate the number of large impacts onto the terrestrial planets. The details of these calculations are described below.

2.5.1　Impact probabilities

In my initial simulations I did not follow any of the particles all the way to impact with any of the planets; particles that entered the inner solar system were stopped either at the Hill sphere of Mars or at an inner boundary of 1 AU heliocentric distance. To compute the impact probabilities for the terrestrial planets, I performed an additional simulation using the results of Sim 1. Only inward-going particles from Sim 1 were considered, as outward-going ones are overwhelmingly likely to collide with or be ejected from the solar system by Jupiter, and as Figs. 2.3 and 2.4a illustrate, there are few observed asteroids beyond 3.4 AU, where outward-going asteroids dominate. I first identified a set of "late" particles from Sim 1 that were removed after 1 Gy at an inner boundary (either at the cutoff at 1 AU or by crossing

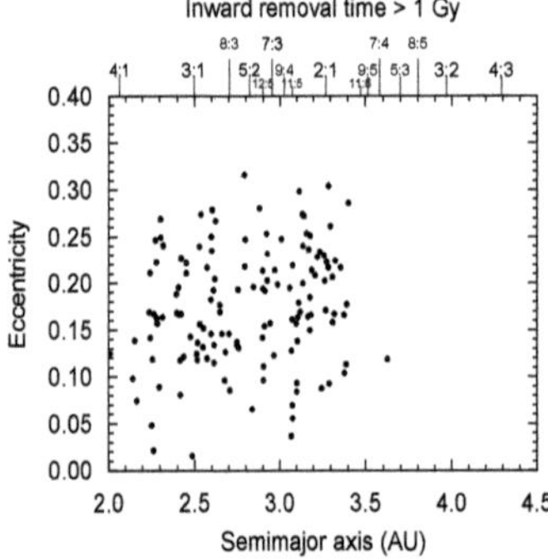
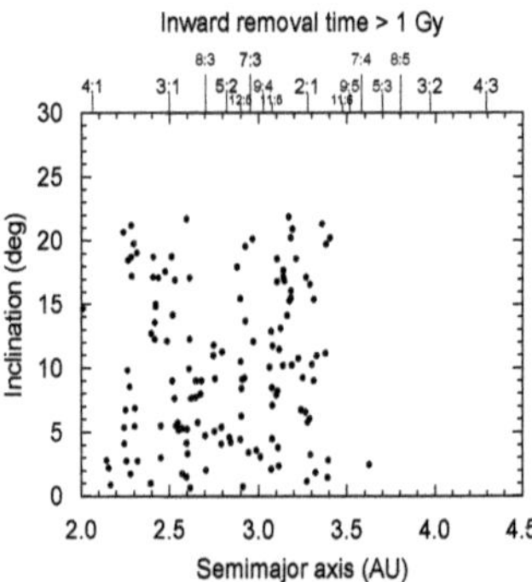

Figure 2.9: Distribution of initial orbital elements of the "late" particles from Sim 1. These particles left the asteroid belt after 1 Gy. Only particles that were removed at an inward-going boundary are shown here, that is they were removed from Sim 1 either by crossing the inner barrier at 1 AU or crossing the Hill sphere of Mars.

the Hill sphere of Mars). The Kirkwood gaps are mostly emptied of asteroids by 1 Gy, as shown in Fig. 2.3. Fig. 2.9 shows the initial semimajor axes, eccentricities, and inclinations of these 136 particles. Note from this figure that asteroids lost due to dynamical erosion after 1 Gy can come from nearly anywhere in the main belt.

First I captured the positions and velocities of the 136 late particles from Sim 1 at the time of their removal. I cloned each of the late particles 128 times, such that $\mathbf{r}_{clone} = (1 + \delta_r)\mathbf{r}_{original}$ and $\mathbf{v}_{clone}(1 + \delta_v)\mathbf{v}_{original}$, where $|\delta_r|, |\delta_v| < 0.001$. The resulting 17408 particles were integrated using the MERCURY integrator with its hybrid symplectic algorithm capable of following particles through close encounters with planets (Chambers, 1999). In this simulation, which I designate Sim 3, all eight major planets were included, the nominal integration step size was 2 days, and an accuracy parameter of 10^{-12} was chosen. The particles were removed if they approached within the physical radius of the Sun or a planet, or if they passed beyond 100 AU. The simulation was run for 200 My. The fraction remaining as a function of time for Sim 3 is shown in Fig. 2.10.

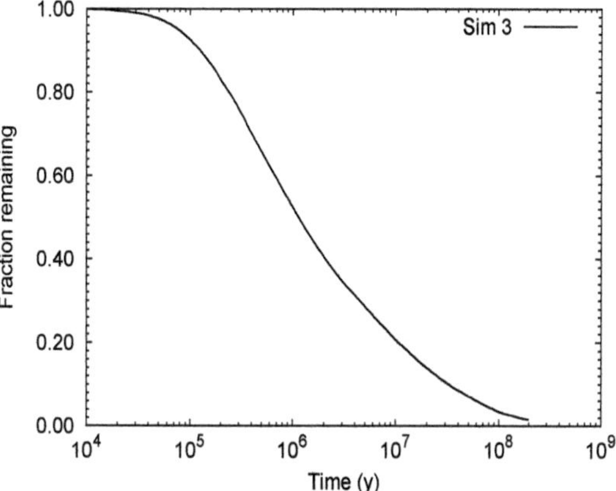

Figure 2.10: Loss history for Sim 3. Particles were lost from this simulation when the impacted a planet, the Sun, or passed beyond 100 AU.

First I captured the positions and velocities of the 136 late particles from Sim 1 at the time of their removal. I cloned each of the late particles 128 times, such that $\mathbf{r}_{clone} = (1 + \delta_r)\mathbf{r}_{original}$ and $\mathbf{v}_{clone}(1 + \delta_v)\mathbf{v}_{original}$, where $|\delta_r|, |\delta_v| < 0.001$. The resulting 17408 particles were integrated using the MERCURY integrator with its hybrid symplectic algorithm capable of following particles through close encounters with planets (Chambers, 1999). In this simulation, which I designate Sim 3, all eight major planets were included, the nominal integration step size was 2 days, and an accuracy parameter of 10^{-12} was chosen. The particles were removed if they approached within the physical radius of the Sun or a planet, or if they passed beyond 100 AU. The simulation was run for 200 My.

For every particle that was removed in Sim 3 (either by impact or escape) I determined from which of the 136 source particles from Sim 1 it was cloned. I used the initial semimajor axis of a given source particle and placed it into a bin of 0.015 AU in width, which I call the source bin. I then weighted the removal event by a factor equal to the ratio of the abundance of observed $H \leq 10.8$ asteroids to the abundance of particles at the end of Sim 1 in the source bin. The weighting factor, which also quantifies the relative amounts of depletion throughout the asteroid belt, is shown in Fig. 2.4b. This weighting accounts for the differences in the orbital distribution of my simulated asteroid belt and the observed large asteroids. The raw impact statistics as well as the probabilities weighted based on the distribution of observed asteroids are tallied in Table 2.2. These give the probability that an asteroid originating in the main belt, and which becomes a terrestrial planet-crosser by dynamical chaos, will impact a planet, the Sun, or be ejected from the solar system. The unweighted and weighted terrestrial impact probabilities for the terrestrial planets differ by $< 10\%$, and both give an impact probability onto the Earth of $\sim 0.3\%$. The weighted impact numbers as a function of time onto the terrestrial planets, the Sun, and the number ejected (removed after crossing the outer boundary at 100 AU) are shown in Fig. 2.11

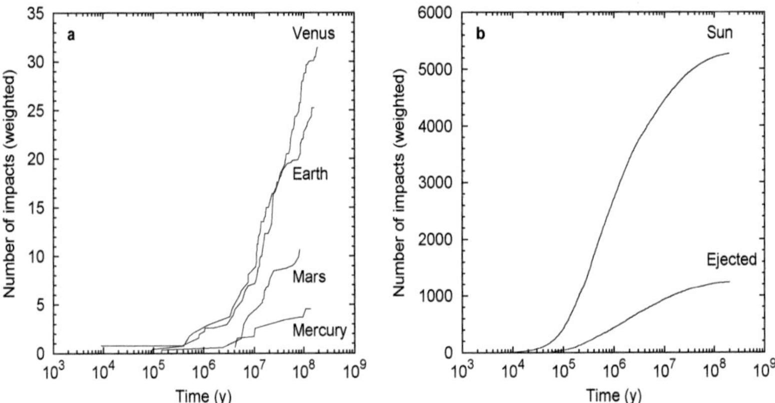

Figure 2.11: Weighted number of loss events from Sim 3.

2.5.2 Flux of large (D>30 km) impactors on the terrestrial planets

I used the weighted impact probabilities shown in Table 2.2 and my model of the loss history of asteroids from Sim 1 to estimate the number of $D > 30$ km impacts onto the terrestrial planets since the end of the LHB. Here I make the assumption that $t = 0$ in my model is about 4 Gy ago, roughly the post-LHB era. Because the loss rate of asteroids is approximately logarithmic, the loss rate is much higher at

Table 2.2: Impact probabilities for the terrestrial planets for Sim 3. Ejected (OSS) refers to particles that either crossed the outer barrier at 100 AU or encountered the Hill sphere of a giant planet.

Fate	Number	%	Weighted %
Ejected (OSS)	13862	78.6	70.0
Survived	269	1.55	13.7
Sun	3111	17.9	15.3
Mercury	10	0.057	0.061
Venus	74	0.425	0.396
Earth	56	0.322	0.306
Mars	26	0.149	0.140

early times than later ones. The early time is likely to have coincided with the tail end of the LHB itself; separating out the component of the impact flux that is due to the LHB rather than to dynamical erosion is problematic. I therefore consider only the dynamical loss at $t > 100$ My in my model as part of the post-LHB epoch. With these assumptions, my model finds that since the end of the LHB, the Earth has experienced ~ 1 impact of a $D > 30$ km asteroid. Venus, with its slightly higher impact probability than Earth, should have experienced ~ 1.3 impacts of this size since the end of the LHB. My model also suggests that Mars and Mercury have had ~ 0.6 and ~ 0.1 impacts of $D > 30$ km asteroids, respectively. These small numbers are consistent with there having been no impacts of $D > 30$ km asteroids on the terrestrial planets since the end of the LHB.

2.5.3 Comparison with record of large impact craters on the terrestrial planets

Known large impact basins on the terrestrial planets are generally of ages confined to the first ~ 800 My of solar system history, the Late Heavy Bombardment (Hartmann, 1965; Ryder, 2002). Excluding those, I consider only the post-LHB large craters on Earth and Venus.

The three largest known impact structures on Earth are Vredefort, Sudbury, and Chicxulub craters, each with final crater diameters $D_c < 300$ km and ages less than ~ 2 Gy (Grieve et al., 2008). Turtle and Pierazzo (1998) argue that Vredefort crater has a diameter of $D_c < 200$ km, which would make the final diameters of all of the largest known impact craters on Earth $D_c \sim 130\text{--}200$ km. At least one of these large impact events has been associated with a mass extinction. The Chixculub crater, estimated to have been created by the impact of a $D \sim 10$ km object, is associated with the terminal Cretaceous mass extinction event (**Alvarez et al.**, 1980; **Hildebrand et al.**, 1991). Estimates from impact risk hazard assessments in the literature suggest that Chixculub-sized impact events happen on Earth on the order of once every 10^8 yr (**Chapman and Morrison**, 1994).

The impact cratering record of Venus is unique in the solar system. Over 98% of the surface of Venus was mapped using synthetic aperture radar by the Magellan spacecraft (**Tanaka et al.**, 1997). The observed craters on Venus appear mostly pristine and are randomly distributed across the planet's surface, which has been taken as evidence for a short-lived, global resurfacing event on the planet within the last ~ 1 Gy (**Phillips et al.**, 1992; **Strom et al.**, 1994). Venus has four craters with $D_c > 150$ km; Mead crater with $D_c = 270$ km, Isabella crater with $D_c = 175$ km, Meitner with $D_c = 149$ km, and Klenova with $D_c = 141$ km.[4] **Shoemaker et al.** (1991) estimated the surface age of Venus to be $\sim 200\text{--}500$ My using the total abundance of Venus craters and a flux of impactors based on the known abundance of Venus-crossing asteroids and on models of their impact probabilities. More recently, **Korycansky and Zahnle** (2005) used similar techniques (as well as an atmospheric screening model for small impactors) and estimated the surface age of Venus to be 730 ± 220 My old. **Phillips et al.** (1992) used four methods to determine the age of Venus' surface, three which used models of observed Venus-crossing asteroids and one that used the observed abundance of craters on the lunar mare as a calibration. All four methods resulted in surface ages between 400-800 My.

Each of the techniques described above for estimating surface ages based on the abundance of observed craters has its shortcomings. Calculating a surface age using the observed population of NEAs makes the assumption that the current population

---

[4]From the USGS/University of Arizona Database of Venus Impact Craters at http://astrogeology.usgs.gov/Projects/VenusImpactCraters/

of NEAs is typical for the entire post-LHB history of the inner solar system, including both the number of near earth asteroids and their computed impact probabilities. Calculating the surface age based on the abundance of craters on the lunar mare makes the assumption that the crater production rate in the inner solar system has been approximately constant over the last ~ 3.2 Gy, which is the age by when most lunar mare were produced (BVSP, 1981).

Large impactors produce craters with a complex morphology. To determine the sizes of impactors which created the largest post-LHB terrestrial impact craters, I used Pi scaling relationships to estimate the transient crater diameter as a function of projectile and target parameters; I then applied a second scaling relationship between transient crater diameter and final crater diameter. The particular form of Pi scaling used here is given by Collins et al. (2005) for impacts into competent rock:

$$D_{tc} = 1.161 \left(\frac{\rho_i}{\rho_t}\right)^{1/3} D^{0.78} v_i^{0.44} g^{-.22} \sin^{1/3}\theta, \tag{2.6}$$

where ρ_i and ρ_t are the densities of the impactor and target in kg m^{-3}, D is the impactor diameter in m, v_i is the impactor velocity in m s^{-1}, g is the acceleration of gravity in m s^{-2} and θ is the impact angle. I used the relationship between final crater diameter, D_c, and transient crater diameter, D_{tc}, given by McKinnon and Schenk (1985):

$$D_c = 1.17 \frac{D_{tc}^{1.13}}{D_*^{0.13}}, \tag{2.7}$$

where D_* is the diameter at which the transition between simple and complex crater morphology occurs. The transition diameter, D_*, is inversely proportional to the surface gravity of the target (Melosh, 1989), and can be computed based on the nominal value for the Moon of $D_{*,moon} = 18$ km.

Applying Eqs. (2.6) and (2.7) to the problem of terrestrial planet cratering by asteroids requires some assumptions about both the impacting asteroids and the targets. For simplicity I assumed that target surfaces have a density $\rho_t = 3$ g cm^{-3} and that impacts occur at the most probable impact angle of 45° (Gilbert, 1893). The characteristic impact velocity is often chosen to be the *rms* impact velocity obtained from a Monte Carlo simulation of planetary projectiles (BVSP, 1981). However, using the *rms* impact velocity may lead to misleadingly high estimates of the impact velocity because the impact velocity distributions are not Gaussian (Bottke et al., 1994; Ito and Malhotra, 2009). Therefore the median velocity may be more appropriate estimate of a "typical" impact velocity (Chyba, 1990, 1991).

I made my own estimates of the impact velocities onto the planets using the results of Sim 3. The small number of impacts computed in Sim 3 makes determining impact velocity distributions difficult. I improved the statistics by using the far more numerous close encounters between planets and test particles to calculate mutual encounter velocities. For every close encounter in Sim 3 I recorded the closest-approach distance and mutual velocity between particles and planets. I only considered particles that had a closest-approach distance within the Hill sphere of a planet. I estimated impact velocities using the *vis viva* integral given by:

$$\frac{1}{2}v_{imp}^2 - \frac{GM_p}{r_p} = \frac{1}{2}v_{enc}^2 - \frac{GM_p}{r_{enc}}, \tag{2.8}$$

where v_{imp} and r_p are the estimated impact velocity and the radius of the planet, and v_{enc} and r_{enc} are the mutual encounter velocity and closest approach distance. Some particles encountered a planet multiple times, which skews the impact velocity estimates, so I only considered unique encounters between a particular particle and a planet. If a particle encountered a planet multiple times it was treated as a single event and I only used the impact velocity of the first encounter. The median, mean, and RMS impact velocities for each of the planets is shown in Table 2.3. The velocity distributions are shown in Fig. 2.12.

Table 2.3: Estimated impact velocities for close encounter events in Sim 3

Planet	Encounters	Velocity (km s^{-1})		
		Median	Mean	RMS
Mercury	3885	38.1	40.5	43.3
Venus	6974	23.4	25.9	27.5
Earth	3522	18.9	20.3	21.1
Mars	1205	12.4	13.1	13.8

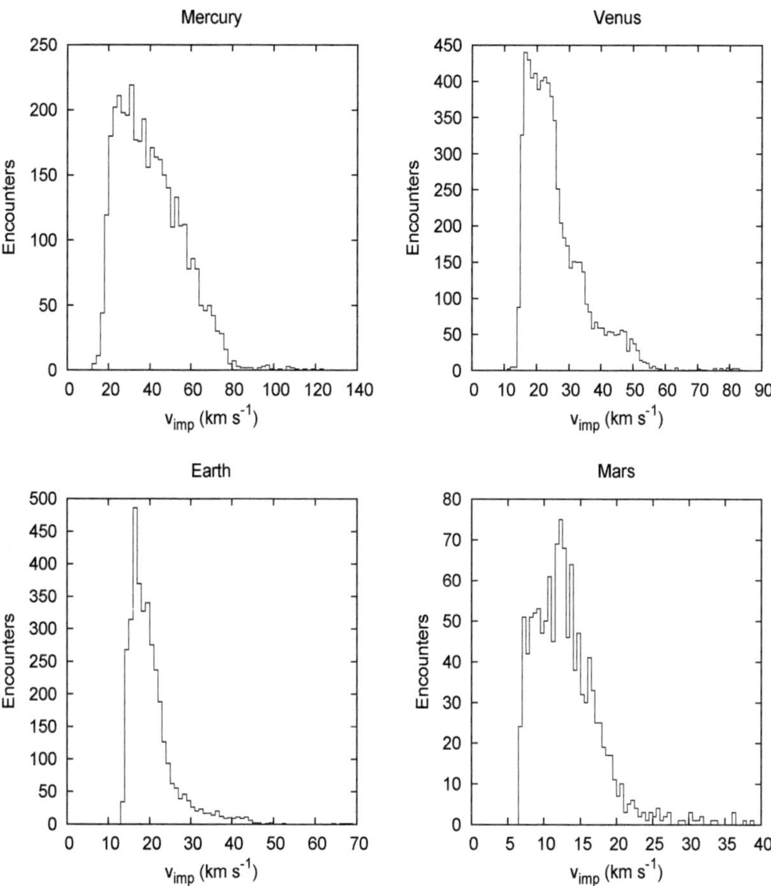

Figure 2.12: Impact velocity distributions of asteroids on the terrestrial planets.

Using the median impact velocities and my assumptions about asteroid density ($\rho_i = 1.5$–3 g cm^{-3}), I calculated the sizes of the impactors that produced the largest post-LHB craters in the inner solar system, using Eqs. (2.6) and (2.7). The "big three" terrestrial impact craters are consistent with asteroidal impactors of diameter $D \sim 7$–14 km. The estimated projectile size for Mead crater, the largest impact crater on Venus, is $D \sim 12$–16 km. Varying the assumptions used in the Pi scaling, Mead and Isabella craters are fully consistent with an impact of a $D > 10$ km asteroid, but Meitner and Klenova craters could be consistent with smaller asteroidal impacts.

In summary, there are no known impact structures of ages $\lesssim 3.8$ Gy (post-LHB) attributed to projectiles with $D \gtrsim 30$ km. This is consistent with the theoretical estimate above. The small number of observed large impacts are consistent with the impact of $D > 10$ km objects. Earth has been impacted by at least 3 objects that are consistent with $D > 10$ km asteroids. During the Phanerozoic eon (the last 545 My), only one impact crater, Chixculub, has been discovered on Earth that is consistent with a $D > 10$ km asteroid. Venus has been impacted by 2–4 $D > 10$ km objects in the last Gy, based on the number of observed large craters and its estimated surface age.

2.5.4   Flux of D>10 km impactors on the terrestrial planets

In my discussion of the dynamical erosion of the main asteroid belt, I have confined myself to $D > 30$ km primordial asteroids because non-gravitational and collisional effects are negligible for this population. However, considering that the largest craters on the terrestrial planets correspond to impactors $D \sim 10$ km, somewhat smaller than 30 km, I am motivated to consider the $D \gtrsim 10$ km population of the main belt. In the size range $D = 10$–30 km the effects of collisions and non-gravitational forces are not negligible, but they do not dominate that of dynamical chaos on the semimajor axis mobility of main belt asteroids. Therefore I extend my dynamical calculations to $D > 10$ km asteroids and I compare the results with the terrestrial planet impact crater record, with the caveat that my results are only a rough estimate of the actual impact flux.

In order to turn the asteroid loss rate into an impact flux for a given crater size, the results of Sim 1, previously normalized to the abundance of $H \leq 10.8$ asteroids (Fig. 2.2), were scaled to the fainter asteroids ($H < 13.2$). Applying $\rho_v = 0.09$ for the geometric albedo, $H < 13.2$ corresponds to $D > 10$ km. The size distribution of the large asteroids of the main belt has not substantially changed over the last ~ 4 Gy and is described well by the present size distribution (**Bottke et al., 2005a; Strom et al., 2005**). I took the $H \leq 10.8$ ($D > 30$ km) loss rate shown in Fig. 2.2 and scaled it to $H < 13.2$ ($D > 10$ km) assuming the asteroid cumulative size distribution in this size range is a simple power law of the form:

$$N_{>D} = kD^{-b}. \tag{2.9}$$

The debiased main belt asteroid size frequency distribution determined by **Bottke et al. (2005a)** gives $b = 2.3$ in the size range 10 km $< D <$ 30 km.

I used my estimate of the loss of $D > 10$ km asteroids from the main belt due to dynamical diffusion, along with my estimate of the impact probabilities shown in Table 2.2, to determine the impact flux of $D > 10$ km asteroids, $\dot{N}_{>10 \text{ km}}$, on the terrestrial planets. The impact flux is based on the four-component piecewise decay law given by Eq. (2.5) with parameters given in Table 2.1, which gives the fraction remaining as a function of time. First the derivative of Eq.(2.5) is taken, yielding $\dot{f}(t)$. In order to convert from a fraction rate, $\dot{f}(t)$, to the flux of $D > 10$ km impacts, $\dot{N}_{>10 \text{ km}}(t)$, I multiplied $\dot{f}(t)$ by the coefficient, $C_{>10 \text{ km}}$, defined as:

$$C_{>10 \text{ km}} = \frac{931}{f(4 \text{ Gy})} \left(\frac{10}{30}\right)^{-2.3} p_i \tag{2.10}$$

The first component of the coefficient is the constant, $931/f(4 \text{ Gy})$, which normalizes the fraction remaining such that the total number of asteroids remaining at $t = 4$ Gy is 931; the latter is the total number of $H < 10.8$ ($D > 30$ km) asteroids in the observational sample. The next component is $(10/30)^{-2.3}$, which scales the results to $D > 10$ km, as given by Eq. (2.9). Finally, the coefficient was multiplied by the weighted probability p_i that terrestrial planet-crossing asteroids impact a given planet. For the Earth, the impact probability is $p_{Earth} = 0.003$ as given in Table 2.2. By multiplying $\dot{f}(t)$ by the coefficient $C_{>10 \text{ km}}$ I obtained the flux of impacts by $D > 10$ km asteroids on the Earth as a function of time, $\dot{N}_{>10 \text{ km}}$. By further multiplying by 1/23, using the ratio of 23:1 Earth:Moon impacts calculated by Ito and Malhotra (2006), I also estimated the lunar flux. The results are shown in Fig. 2.13a, showing my estimate of the impact flux of $D > 10$ km asteroids on the Earth (left-hand axis) and Moon (right-hand axis) since 1 My after the establishment of the current dynamical architecture of the main asteroid belt.

For comparison, I also plot in Fig. 2.13a the impact flux estimates obtained by Neukum et al. (2001) from crater counting statistics. The shaded region of Fig. 2.13a represent upper and lower bounds on the estimated post-LHB impact flux using calibrated lunar cratering statistics (Neukum et al., 2001). The lower bound is calculated using the number of $D_c > 200$ km lunar impact craters from the Neukum Production Function (NPF, Fig. 2 of Neukum et al. (2001)); the rate is 7×10^{-9} craters km^{-2} Gy^{-1}. The upper bound is calculated using the number of $D_c > 140$ km lunar impact craters from the NPF; the rate is 2×10^{-8} craters km^{-2} Gy^{-1}. These rates were multiplied by the surface area of the Moon to obtain the number of impacts on the lunar surface, and then multiplied by 23 to obtain the number of impacts on the Earth's surface.

I have also calculated the cumulative number of impacts on a surface with a given age; the result is shown in Fig. 2.13b. The cumulative number of impacts as a function of time is calculated simply as $N_{cumulative} = C_{>10 \text{ km}} \left[f(t) - f(4 \text{ Gy}) \right]$. Assuming that $t = 0$ corresponds with an age of 4 Gy ago, the surface age is simply $SA = 4 \text{ Gy} - t$. The upper and lower bounds on the cumulative number of impacts calculated from the NPF are shown as the shaded region, similar to Fig. 2.13a. My calculations of the impact flux show that at early times the flux was larger than the estimate given by the NPF, but that after $t = 500$ My the flux of impacts that I calculate is lower than that given by the NPF. The present day flux of $D > 10$ km impactors is currently about an order of magnitude lower than that given by the NPF.

I note that the dynamical erosion rate decays with time, and the impact rate that I derive also decays as a function of time. My estimated impact flux for large craters declines by a factor of 3 over the last 3 Gy, as shown in Fig. 2.13. This is in some contrast with previous studies: in commonly used cratering chronologies, the impact flux is usually assumed to have been relatively constant over time since the end of the LHB. The results obtained for the Earth-Moon system shown in Fig. 2.13 can also be applied to the remaining terrestrial planets, using the weighted impact probabilities, p_i, given in Table 2.2. My estimated decay of the large impact rate for Mars over the last 4 Gy agrees with the results of **Quantin et al. (2007)** that suggest that the impact cratering rate of $D_c > 1$ km craters on Mars has declined by a factor of 3 over the last 3 Gy, based on counts of craters on 56 landslides along the walls of Valles Marineris. Reliable estimates are lacking for the absolute ages of lunar surfaces with ages < 3 Gy, but what estimates do exist (i.e. estimated ages of Copernicus, Tycho, North Ray, and Cone craters from the Apollo missions) seem to be mostly consistent with a constant flux of impactors after 3 Gy ago, but with a possible increase in the flux after 1 Gy ago (**Stöffler and Ryder, 2001**). Also, estimates of the current impact flux from studies of the lunar record seem to be consistent with estimates made by observing the modern NEO population and estimating their impact probabilities (**Morrison et al., 2002**). However, **Culler et al. (2000)** suggested that the lunar impact flux declined by a factor of 2 or 3 from ~ 3.5 Gy ago until it reached a low at ~ 600 My ago and then increased again, based on dating of lunar impact glasses in soils. This hypothesis is consistent with my result, that the overall impact flux onto the terrestrial planets has decreased by a factor of 3 since ~ 4 Gy ago due to the dynamical erosion of the asteroid belt. The apparent increase in the flux at ~ 600 My ago until the present may be the result of a few large asteroid breakup events in the inner asteroid belt, such as the Flora and Baptistina family-forming events, and therefore the modern NEO population may not be representative of the NEO population over the last ~ 4 Gy.

Assuming that $t = 0$ corresponds to an age of 4 Gy ago, I estimate that in the last 3 Gy there have been ~ 4 impacts of $D > 10$ km asteroids on the Earth due

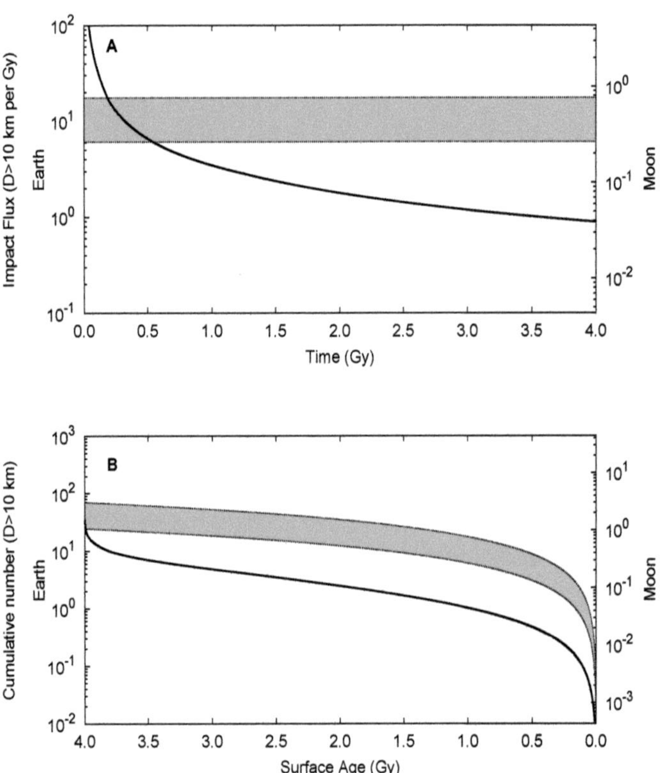

Figure 2.13: Impact rates of $D > 10$ km asteroids on the Earth and Moon. **a)** Solid line is the impact flux of $D > 10$ km impactors per Gy onto the Earth and the Moon, from our calculations. **b)** Solid line is the cumulative number $D > 10$ km impactors for a given surface age, from our calculations. The shaded regions indicate the Neukum Production Function (**Neukum et al., 2001**) upper and lower bounds on the production of $D_c > 140$–200 km impact craters.

to dynamical erosion, and in the last 1 Gy there has been only 1 impact. This is about an order of magnitude lower than estimates of the cratering rate of the terrestrial planets using the lunar cratering record, which give ~ 6–18 $D > 10$ km impactors onto the Earth every 1 Gy using the NPF and assuming a $D > 10$ km object produces a 150–200 km crater (Neukum et al., 2001; Chapman and Morrison, 1994). The discrepancy between my estimates of the production of large craters and those based on the NPF may indicate that either (1) the current production rate of large craters ($D_c \gtrsim 150$ km) is substantially overestimated by the NPF, or (2) the current production rate of large craters is not dominated by chaotic transport of large main belt asteroids.

One possibility is that cratering at this size is dominated by comets, rather than asteroids. The fraction of terrestrial planet impacts that are due to comets vs. asteroids has long been controversial, but is generally thought that Jupiter-family comets contribute fewer than 10% and Oort cloud comets contribute fewer than 1%. (Bottke et al., 2002b; Stokes et al., 2003). Even periodic comet showers may not substantially increase the impact contribution from comets (Kaib and Quinn, 2009). Therefore it seems unlikely that comet impacts can account for an order of magnitude more large impacts on the terrestrial planets than asteroids.

Another possibility is that large asteroid breakup events (followed by fragment transport via the Yarkovsky effect to resonances) dominate the production of large craters. My model neglects breakup events which have produced numerous $D \sim$ 10 km fragments over the last 4 Gy, and breakup events near resonances with Earth-impact probabilities higher than that of the intrinsic main belt may contribute to the large basin impact rate. This is similar to the hypothesis proposed by Bottke et al. (2007b) for the origin of the Chicxulub impactor from the Baptistina family-forming event. The Baptistina breakup is hypothesized to have involved two large asteroids ($D_1 \sim 170$ km and $D_2 \sim 60$ km) that collided at a semimajor axis distance $<$ 0.01 AU from two overlapping weak resonances that, combined, increased fragment eccentricities to planet-crossing orbits with a 1.7% Earth-impact probability (Bottke et al., 2007b). It is unclear whether such fortuitous combinations of conditions occur often enough to dominate the production of large craters over the solar system's history since the end of the LHB. Large family forming breakup events do occur in the asteroid belt, and they likely can increase the flux of impacts onto the terrestrial planets (Nesvorný et al., 2002, 2006; Korochantseva et al., 2007). The Flora family breakup event in particular was large and occurred in the inner asteroid belt roughly 0.5–1 Gy ago (Nesvorný et al., 2002). Also, while the Yarkovsky effect is very weak for $D > 10$ km asteroids, it is not entirely negligible over the age of the solar system. For instance, the loss rate of $D = 10$ km from the inner asteroid belt is somewhat higher when the Yarkovsky effect is taken into account compared to when it is not (Bottke et al., 2002a). It is doubtful whether this modest difference can account for a factor of ten increase in the flux of $D > 10$ km objects in the terrestrial planet region, however additional modeling may be needed to confirm this. In addition, the weakness of the Yarkovsky effect on large asteroids can paradoxically enhance their mobility in the inner main belt. Combinations of nonlinear secular resonances and weak three-body resonances may case asteroids to slowly diffuse through the middle and inner main belt (Carruba et al., 2005; Michtchenko et al., 2009). Only large asteroids that are only weakly affected by the Yarkovsky effect can remain inside these resonances long enough for them to act.

Finally, I consider the contribution from high velocity impactors with $D < 10$ km to the large impact crater production rate. The velocity distributions of asteroid impacts on the terrestrial planets have significant high velocity tails, as seen in Fig. 2.12. Because size distributions of asteroids follow a power law with a negative index, as in Eq. (2.9), smaller objects are more numerous than larger ones. I calculated the relative contribution of impacts by objects of varying sizes on the production of craters of a given size. Using Eqs. (2.6) and (2.7), a $D = 10$ km projectile striking the Earth at 19 km s^{-1} produces a crater with a final diameter $D_c = 187$ km, assuming target and projectile densities are both 3 g cm^{-3}, and the impact occurs at a 45° angle. I solved for the projectile size needed to produce a $D_c = 187$ km crater while varying the impact velocity. The result was used to transform the Earth impact velocity distribution from Fig. 2.12 into a projectile

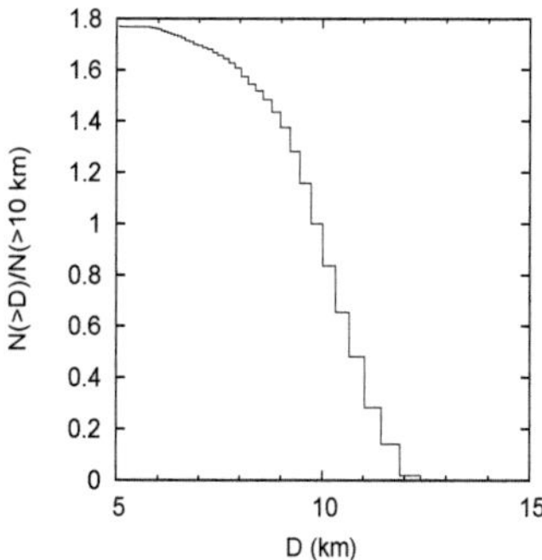

Figure 2.14: Cumulative size distribution of projectiles that contribute to the production of $D_c = 187$ km craters on Earth. This is calculated by solving for the projectile size needed to produce a crater on Earth with a final diameter $D_c = 187$ km for the velocity distribution of Fig. 2.12c, and convolving the result with the asteroid size frequency distribution (Eq. (2.9), with index $b = 2.3$). This cumulative distribution is normalized such that the number of craters produced by objects of diameter $D > 10$ km is unity.

size distribution for a fixed final crater diameter. Finally I multiplied the binned projectile size distribution by a binned asteroid size distribution, using a cumulative distribution as in Eq. (2.9) with an index $b = 2.3$. The result was then turned into a cumulative distribution and normalized to $N > 10$ km, and is plotted in Fig. 2.14. This plot shows that asteroids with $D < 10$ km that impact at high velocity increase the production of large craters by no more than a factor of two. Therefore $D < 10$ km asteroids impacting at high velocity cannot account for the order of magnitude difference in the production rate of large impact craters on the Earth between my model and the NPF.

On very ancient terrains, the discrepancy between total number of accumulated craters estimated from my model compared with the constant-flux models used in crater chronologies is less than with younger surfaces, as illustrated by Fig. 2.13b. My model also cannot account for the LHB itself. Total accumulated craters on ancient heavily cratered terrains associated with the LHB are at least 10–15 times higher than on younger terrains (Stöffler and Ryder, 2001), and likely even more if the surfaces reached equilibrium cratering. The impact flux during the LHB was at least two orders of magnitude higher than the average flux over the last 3.5 Gy, and possibly three orders of magnitude more if the LHB was a short-lived event (Ryder, 2002). Even with the more enhanced rate of impacts at early times, and assuming higher Earth impact probabilities of 3% based on estimates from the ν_6 resonance, the impact flux due to dynamical erosion is roughly one or two orders of magnitude lower than that needed to produce the heavily cratered terrains associated with the LHB (Stöffler and Ryder, 2001; Neukum et al., 2001).

2.6   Conclusion

The main asteroid belt has unstable zones associated with strong orbital resonances with Jupiter and Saturn and relatively stable zones elsewhere. In most of the strongly unstable zones, the timescale for asteroid removal is $\lesssim 1$ Myr; one exception is the 2:1 mean motion resonance with Jupiter where the timescale to empty this resonance approaches a gigayear. The relatively stable zones also lose asteroids by means of weak dynamical chaos on long timescales. I have found that the dynamical loss of test particles from the main asteroid belt as a whole is best described as a piecewise logarithmic decay. A piecewise logarithmic decay law implies that the intrinsic loss rate from the asteroid belt decays inversely proportional to time, $\dot{n} \propto t^{-1}$, but with different proportionality constants for different regions inside the belt. When a region with a particular decay rate empties of asteroids, the decay law for the entire asteroid belt undergoes a change in slope. This logarithmic loss of asteroids due to dynamical chaos was established very soon after the current dynamical architecture of the asteroid belt was established, and it continues to the present day with very little deviation. Dynamical chaos is the predominant mechanism for the loss of large asteroids, $D \gtrsim 10$–30 km. I have calculated that the asteroid belt 1 My after the establishment of the current dynamical architecture of the solar system had roughly twice its present number of large asteroids. Because their loss rate is inversely proportional to time, I estimate that the flux of large impactors has declined by a factor of 3 over the last 3 Gy. I have calculated that large asteroidal impactors originating across the main asteroid belt have an overall Earth impact probability of 0.3%, and that the number of impacts of $D > 10$ km asteroids on Earth is only ~ 1. My result on the current impact flux of $D > 10$ km asteroids due to dynamical erosion of the asteroid belt is an order of magnitude less than the values adopted in the recent literature on crater chronology and impact hazard assessment. I have evaluated several possible explanations for the discrepancy and find them inadequate. My results can be used to improve studies of large impacts on the terrestrial planets.

CHAPTER 3

A RECORD OF PLANET MIGRATION IN THE ASTEROID BELT

Portions of the content of this chapter has been previously published in the journal *Nature* as **Minton and Malhotra (2009)**.

3.1 Introduction

The main asteroid belt lies between the orbits of Mars and Jupiter, but the region is not uniformly filled with asteroids. There are gaps, known as the Kirkwood gaps, in the asteroid distribution in distinct locations that are associated with orbital resonances with the giant planets (**Kirkwood, 1867**); asteroids placed in these locations will follow chaotic orbits and be removed (**Wisdom, 1987**). Here I show that the observed distribution of main belt asteroids does not fill uniformly even those regions that are dynamically stable over the age of the solar system. I find a pattern of excess depletion of asteroids, particularly just outward of the Kirkwood Gaps associated with the 5:2, the 7:3, and the 2:1 jovian resonances. These features are not accounted for by planetary perturbations in the current structure of the solar system, but are consistent with dynamical ejection of asteroids by the sweeping of gravitational resonances during the migration of Jupiter and Saturn ~ 4 Gy ago.

The Kirkwood gaps have been explained by the perturbing effects of the giant planets that cause dynamical chaos and orbital instabilities on very long timescales in narrow zones in the main asteroid belt (**Wisdom, 1987**), but thus far it has not been established how much of the main belt asteroid distribution is accounted for by planetary perturbations alone. I compared the distribution of observed asteroids against a model asteroid belt uniformly populated in the dynamically stable zones. My model asteroid belt was constructed as follows. Test particle asteroids were given eccentricity and inclination distribution similar to the observed main belt, but a uniform distribution in semimajor axis. I then performed a numerical integration for 4 Gy of the test particles' orbital evolution under the gravitational perturbations of the planets using a parallelized implementation of a symplectic mapping (Wisdom and Holman, 1991; Saha and Tremaine, 1992).

3.2   Asteroid belt model

I will use the model asteroid belt containing 5760 test particles integrated for 4 Gy described in Chapter 2 (Sim 1). I sorted the surviving particles into semimajor axis bins of width 0.03 AU. I compared the model asteroid belt with the observed asteroid belt, as shown in Fig. 3.1a. I find that the observed asteroid belt is overall more depleted than the model can account for, and there is a particular pattern in the excess depletion: there is enhanced depletion just exterior to the major Kirkwood gaps associated with the 5:2, 7:3, and 2:1 mean motion resonances (MMRs) with Jupiter (the regions spanning 2.81–3.11 AU and 3.34–3.47 AU in Fig. 3.1a); the regions just interior to the 5:2 and the 2:1 resonances do not show significant depletion (the regions spanning 2.72–2.81 AU and 3.11–3.23 AU in Fig. 3.1a), but the inner belt region (spanning 2.21–2.72 AU) shows excess depletion.

The above conclusions about the patterns of depletion are based on my model asteroid belt which assumes uniform initial population of the dynamically stable zones. It is conceivable that the discrepancies between the model and the observations could be due to a non-uniform initial distribution of asteroids. However, the particular features I find cannot be explained by appealing to the primordial distribution of planetesimals in the solar nebula, nor to the effects of the mass depletion that occurred during the planet formation era (see Supplementary Information). As I show below, they can instead be readily accounted for by the effects of giant planet migration in the early history of the solar system.

There is evidence in the outer solar system that the giant planets – Jupiter, Saturn, Uranus and Neptune – did not form where we find them today. The orbit of Pluto and other Kuiper Belt Objects (KBOs) that are trapped in mean motion

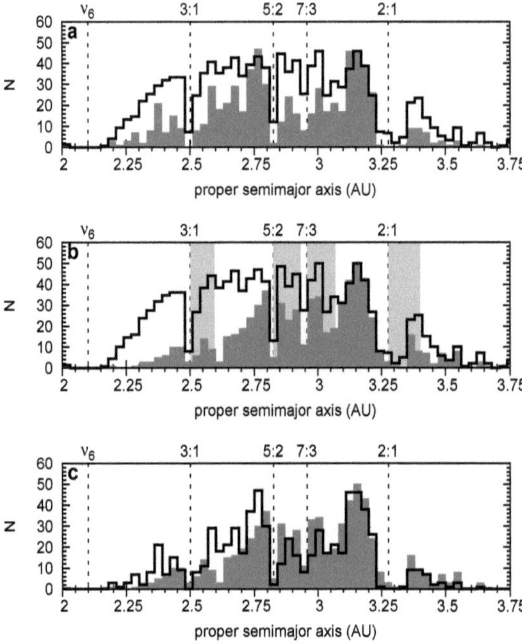

Figure 3.1: Comparison of the observed main belt asteroid distribution with my simulated asteroid belt and results of the migration simulation. **a**, The solid line histogram is the distribution of asteroids remaining in my model asteroid belt at the end of the 4 Gy simulation in which the asteroid belt region was initially uniformly populated with test particles and the planets were in their current orbits. The shaded histogram is my observational comparison sample. The model asteroid belt (solid line) was normalized by multiplying all bins by a constant such that the value of the most-populous model bin equaled that of its corresponding bin in the observations. The current positions of the ν_6 secular resonance and the strong jovian mean motion resonances associated with the major Kirkwood gaps are shown. **b**, The solid line is the initial distribution of test particles in the simulation with migrating planets. The shaded histogram is the normalized distribution of test particles remaining at the end of the 100 My migration simulation. The planet migration history followed the form of Eq. 3.1. The grey-shading indicates regions swept by the strong jovian mean motion resonances. **c**, Comparison of the model asteroid belt subjected to planet migration and the observed asteroid belt. The solid line is the distribution of observed large asteroids.The shaded histogram is the distribution of test particles remaining at the end of the 100 My migration simulation.

resonances with Neptune can be explained by the outward migration of Neptune due to interactions with a more massive primordial planetesimal disk in the outer regions of the solar system (Malhotra, 1993, 1995). The exchange of angular momentum between planetesimals and the four giant planets caused the orbital migration of the giant planets until the outer planetesimal disk was depleted of most of its mass, leaving the giant planets in their present orbits (Fernandez and Ip, 1984; Hahn and Malhotra, 1999; Tsiganis et al., 2005). As Jupiter and Saturn migrated, the locations of mean motion and secular resonances swept across the asteroid belt, exciting asteroids into terrestrial planet-crossing orbits, thereby greatly depleting the asteroid belt population and perhaps also causing a late heavy bombardment in the inner solar system (Liou and Malhotra, 1997; Levison et al., 2001; Gomes et al., 2005; Strom et al., 2005).

3.3   Simulation of planet migration and its effects on the asteroid belt

I performed a computer simulation to test the hypothesis that the patterns of asteroid depletion inferred from Fig. **3.1**a are consistent with planet migration. I used a total of 1819 surviving test particles from the previous 4 Gy simulation as initial conditions for a simulation with migrating planets. For the purposes of this simulation, I applied an external tangential force on each of the planets to simulate their orbital migration, so that a planet's semimajor evolved as follows (Malhotra, 1993):

$$a(t) = a_0 + \Delta a \left[1 - \exp\left(-t/\tau\right)\right], \tag{3.1}$$

where a_0 is the initial semimajor axis, Δa is the migration distance, and τ is a migration rate e-folding time. Jupiter, Saturn, Uranus, and Neptune had initial semimajor axes displaced from their current values by $\Delta a = +0.2, -0.8, -3.0$, and -7.0 AU, respectively; these values are consistent with other estimates of Jupiter's and Neptune's migration distances (Fernandez and Ip, 1984; Malhotra, 1995; Franklin et al., 2004; Tsiganis et al., 2005), but Uranus' and Saturn's migration distances are less certain. I used $\tau = 0.5$ My, which is near the lower limit inferred from Kuiper belt studies (Murray-Clay and Chiang, 2005). After 100 My of evolution under the influence of migrating planets, the 687 surviving test particles in the simulation were sorted and binned. The distribution of the survivors is shown in Fig. 3.1b.

In contrast with Fig. 3.1a, the asteroid belt distribution produced by the planet migration model matches qualitatively quite well the distribution of the observed asteroids (Fig. 3.1c). The depletion patterns in Fig. 3.1a may be attributed to a number of plausible causes. Any mechanism invoked to explain the depletion must account for the features seen in Fig. 3.1a, namely that there is enhanced depletion in regions just outward of the major Kirkwood gaps. Here I explain why these features cannot be accounted for by appealing to the primordial distribution of planetesimals in the solar nebula, or to the effects of the first mass depletion event that occurred during the planet formation era.

3.4   Discussion

First, I consider whether the depletion may be a reflection of the primordial mass distribution of planetesimals in the solar nebula, which is related to the surface mass density of the nebular disk, usually expressed as a decreasing function of heliocentric distance, $\Sigma \propto r^{-p}$. The "minimum mass solar nebula" (MMSN) model, which is derived by spreading out the masses of each planet in an annulus centered on its presumed initial semimajor axis and then fitting a surface mass density function, estimates that the index $p \approx 1.5$ if the giant planets are assumed to have formed in their initial locations, or $p \approx 2.2$ if the giant planets are assumed to have formed in a more compact configuration (Weidenschilling, 1977; Hayashi, 1981; Desch, 2007). If mass is equally distributed among planetesimals of similar size in the asteroid belt region, then the number distribution of planetesimals is $N \propto r^{-p+1}$. Fig. 3.1a indicates that in the region between the inner edge of the asteroid belt and the 5:2 resonance, the number distribution of asteroids within the dynamically stable regions is an increasing function of heliocentric distance; neglecting the enhanced depletion adjacent to the Kirkwood gaps (the shaded regions in Fig. 3.1b), the overall depletion is approximately linear with r in the region from 2.25 AU to 2.8 AU, implying that $N \propto r$. Therefore a nebular surface mass density in accord with the MMSN model is inconsistent with that required to produce the observed asteroid distribution, and furthermore cannot account for the enhanced depletion adjacent to the Kirkwood gaps.

Second, I consider whether the asteroid distribution may have been produced during the planet formation era, due to either secular resonance sweeping related to the depletion of the solar nebula (Heppenheimer, 1980), or by gravitational perturbations from embedded planetary embryos in the asteroid belt region (Wetherill, 1992). The sweeping secular resonance model may account for depletion in the inner asteroid belt seen in Fig. 3.1a if the rate of sweeping was an increasing function of semimajor axis. In the embedded planetary embryo model, the depletion in the inner asteroid belt could be accounted for if the scattering process was more efficient near the inner edge with decreasing efficiency outward. Thus the overall depletion trend does not contradict either model. However, neither model can readily account for the observed enhanced depletion adjacent to the Kirkwood gaps.

Of note is that the inner asteroid belt region (2.15–2.81 AU) is somewhat more depleted in the migration simulation than in the observations. The majority of depletion from this region found in my migration simulation is due to the sweeping ν_6 secular resonance. This powerful resonance removes asteroids from the main belt by secularly increasing their eccentricities to planet-crossing values (Murray and Dermott, 1999). The maximum eccentricity of an asteroid disturbed by the passage of the ν_6, and thereby the degree of asteroid depletion, is related to the sweeping speed: the slower the sweeping the more the depletion (Heppenheimer, 1980). The distances the planets migrate determine the ranges in asteroids' semimajor axes that are affected by the sweeping. As I will show in Chapter 4, in my simulation, as Jupiter and Saturn migrated, the ν_6 secular resonance swept inward across the entire main asteroid belt to its present location at ~ 2.1 AU, as shown in Fig. 4.3. But because the ν_6 resonance location is such a steep function of Saturn's semimajor axis (see Fig. 4.3), for even modest proposed values of Saturn's migration distance, all of the asteroid belt is affected by the passage of this resonance. Thus, the overall level of depletion of the asteroid belt is most strongly dependant on the speed of planet migration, and only secondarily on the migration distance. Because I used an exponentially decaying migration rate for the giant planets, the ν_6 resonance sweeping rate decreased as it approached its current location, thereby causing relatively greater asteroid depletion in the inner belt. Thus, the small but noticeable differences between the model and the observations in Fig. **3.1c** are sensitive to the details of the time history of the planet migration speed.

I note that my model asteroid belt lost 62% of its initial pre-migration population, but the actual asteroid belt may have lost as much as $\sim$ 90–95% of asteroids due to migration (**O'Brien et al., 2007**). Because the overall level of asteroid depletion is particularly sensitive to the speed of planet migration, detailed exploration of parameters of the planet migration model and comparison with observations of main belt asteroids may provide strong quantitative constraints on planet migration.

CHAPTER 4

THE LOCATION OF THE ν_6 RESONANCE DURING PLANET MIGRATION

4.1 Introduction

Secular resonances play an important role in the evolution of the main asteroid belt. **Williams (1971)** first noted that the main asteroid belt seemed to be depleted at $a < 2.5$, which was near the secular resonance where $g - g_6$, where g is the rate of precession of the longitude of pericenter, ϖ, of the asteroid and g_6 is the sixth eigenfrequency of the solar system (corresponding roughly to the rate of precession of Saturn's longitude of pericenter). The secular resonance involving $g_0 - g_6 = 0$ (where g_0 is the frequency of precession of the asteroid and g_6 is the sixth fundamental eigenmode of the solar system), also called the ν_6 resonance, is an important resonance for the delivery of NEAs to the inner solar system **Scholl and Froeschle (1991)**. **Williams and Faulkner (1981)** showed that the location of the ν_6 resonance actually form surfaces in $a - e - \sin i$ space, and **Milani and Knezevic (1990)** showed that those surfaces seem to correspond to the "inner edge" of the main asteroid belt. The location of the ν_6 resonance is defined as the location where the rate of precession of a massless particle (or asteroid) is equal to the g_6 eigenfrequency of the solar system (**Murray and Dermott, 1999**).

There is evidence that the solar system was not always in the configuration that we find it today. At some point within the first ~ 700 My of solar system history, the giant planets (Jupiter, Saturn, Uranus, and eptune) may have experienced planetesimal-driven migration (**Fernandez and Ip, 1984**). Evidence for migration of the outer giant planets comes from observations of the structure of the Kuiper belt, an icy belt of planetesimals beyond Neptune (**Malhotra, 1993, 1995; Hahn and Malhotra, 1999; Levison et al., 2008**). The giant planet migration may also have affected the main asteroid belt as both mean motion and secular resonances would have swept through the asteroid belt region and removed asteroids by exciting them into planet-crossing orbits, while also capturing asteroids into the stable 3:2 resonance (Liou and Malhotra, 1997; Franklin et al., 2004; Minton and Malhotra, 2009). The disruption of the asteroid belt through resonance sweeping has been implicated in the so-called "Late Heavy Bombardment," also known as the "Late Lunar Cataclysm," that is thought to have occurred ~ 3.9 Gy ago (Tera et al., 1974; Ryder, 2002; Chapman et al., 2007; Strom et al., 2005; Gomes et al., 2005).

Constraints on the timing and duration of the planet migration from observations may help to constrain models of our early solar system's history. An example of such a constraint from orbital dynamnics is in the distribution of the population of Kuiper Belt Objects that are trapped in the 2:1 resonance with Neptune. Objects in the 2:1 resonance with Neptune can librate about one of two centers, and the speed of migration determines whether or not the libration centers are populated equally; faster migration timescales result in asymmetries in the capture probabilities between the two libration centers (Murray-Clay and Chiang, 2005). In the main asteroid belt, an important mechanism for depleting asteroids during planet migration is the sweeping of the ν_6 secular resonance. In most migration models, Saturn is thought to have originally resided closer to the Sun than currently, and then experienced outward migration Fernandez and Ip (1984); Hahn and Malhotra (1999); Tsiganis et al. (2005). During outward migration, the location of the ν_6 resonance would have swept inwards, passing through the asteroid belt to finally reside in its current location at ~ 2.1 AU (for an asteroid with $i \approx 0$) (Minton and Malhotra, 2009). As the timescale for exciting an asteroid in the ν_6 resonance to planet-crossing values is ~ 1 My (Williams and Faulkner, 1981), the fact that many asteroids apparently survived the sweeping of the ν_6 places some constraints on the timescale of planet migration.

Understanding how to calculate these migration timescale constraints requires an understanding of how the location of the ν_6 resonance varies with the positions of the giant planets. The ν_6 resonance is most sensitive to the semimajor axis of Saturn, as the g_6 frequency is approximately the precession rate of Saturn's longitude of perihelion (Malhotra, 1998). While Minton and Malhotra (2009) estimated the position of the zero-inclination location of the ν_6 as a function of Saturn's semi-major axis (see their Fig. 2), their method was limited and the results are only an approximation. Here I produce a more accurate model of how the zero-inclination location of the ν_6 resonance changes as a function of Jupiter and Saturn's semimajor axis under reasonable scenarios of planet migration.

4.2 Spectral analysis of orbital evolution

The secular variations of planet orbital elements can be modeled as a linear combination of periodic terms. In this work I consider a planar solar system consisting of Jupiter and Saturn only. In this model, the secular variation in eccentricity of planet j is given by:

$$\{h_j, k_j\} = \sum_i E_j^{(i)} \{\cos(g_i t + \beta_i), \sin(g_i t + \beta_i)\}, \tag{4.1}$$

where g_i are the eigenfrquencies of the system and $E_j^{(i)}$ are the corresponding eigenvectors. The solution for a co-planar two-planet solar system has two modes with frequencies designated g_5 and g_6, using the nomenclature of Murray and Dermott (1999). Such a solution places the location of the ν_6 at ~ 1.8 AU, a difference of ~ 0.3 AU from the actual value of 2.1 AU. This difference is due to the proximity of Jupiter and Saturn to the 5:2 mean motion resonance, which alters value of the g_6 eigenfrequency and introduces additional frequencies, such as $g_{10} = 2g_6 - g_5$.

I calculate values for the eccentricity-pericenter eigenfrequencies by direct numerical integration of the two-planet, planar solar system. A Fourier transform of the time history of the $\{h_j, k_j\}$ vector yields the fundamental frequencies. For regular (non-chaotic) orbits, the spectral frequencies are well defined and correspond to the eigenfrequencies of the secular solution. For chaotic orbits, the spectral frequencies the power spectrum does not have peaks at discrete frequencies. Michtchenko and Ferraz-Mello (2001) used spectral analysis of long-term numerical integrations to show that modest displacements in the giant planets' semimajor axes and eccentricities could lead to chaotic orbital evolution and exhibit broadband power spectra. A striking feature of our solar system noted by Michtchenko and Ferraz-Mello is that the particular combination of semimajor axis and eccentricity combinations of each of the four giant planets seem to lie in regions of regular orbits but very near regions of chaos. However, even systems whose power spectra is strongly peaked at small numbers of discrete frequencies can exhibit weakly chaotic behavior. Hayes (2008) showed that changes in the initial conditions of the giant planets as small as one part in 10^7 could lead to either chaotic or regular orbits in numerical integrations, but much larger changes in the positions of the planets were needed by Michtchenko and Ferraz-Mello to change from the strongly peaked spectra to the "noisy" broadband spectra of strongly chaotic orbital evolution.

Here I use the spectral analysis of long-term integrations of the giant planets to investigate the secular properties of the solar system during the period of giant planet migration. I performed 255 numerical simulations of the Jupiter-Saturn pair for ~ 104 My each with a stepsize of 0.1 yr, varying Saturn's semimajor axis between 7.4–10.5 AU while keeping Jupiter fixed at 5.2 AU. Jupiter and Saturn's inclinations were set to zero, but their initial eccentricities their modern values of ~ 0.05. The initial longitude of pericenter and mean anomalies of Jupiter were $\varpi_{jup,i} = 15°$ and $\lambda_{jup,i} = 92°$, and Saturn were $\varpi_{sat,i} = 338°$ and $\lambda_{sat,i} = 62.5°$. After each simulation, I Fourier-decomposed the resultant $\{h, k\}$ vector time series as in Eq. (4.1) to obtain the e–ϖ power spectrum. I then identify the frequencies that are the equivalents to the g_5 and g_6 modes of the solar system. Fig. 4.1 shows examples of some of the spectra obtained over a frequency range of 1–1000″ yr^{-1}. Each panel in Fig. 4.1 is labeled with the mean semimajor axis of Saturn, and the panels are given in order from largest to smallest semimajor axis. The vertical scale in the spectrum plots is arbitrary, and each spectrum is scaled to accommodate the highest amplitude peaks. Table 4.1 lists the initial semimajor axis of Saturn along with the mean value of the semimajor axes of Saturn and Jupiter, and the corresponding g_5 and g_6 frequencies obtained using the fourier spectra.

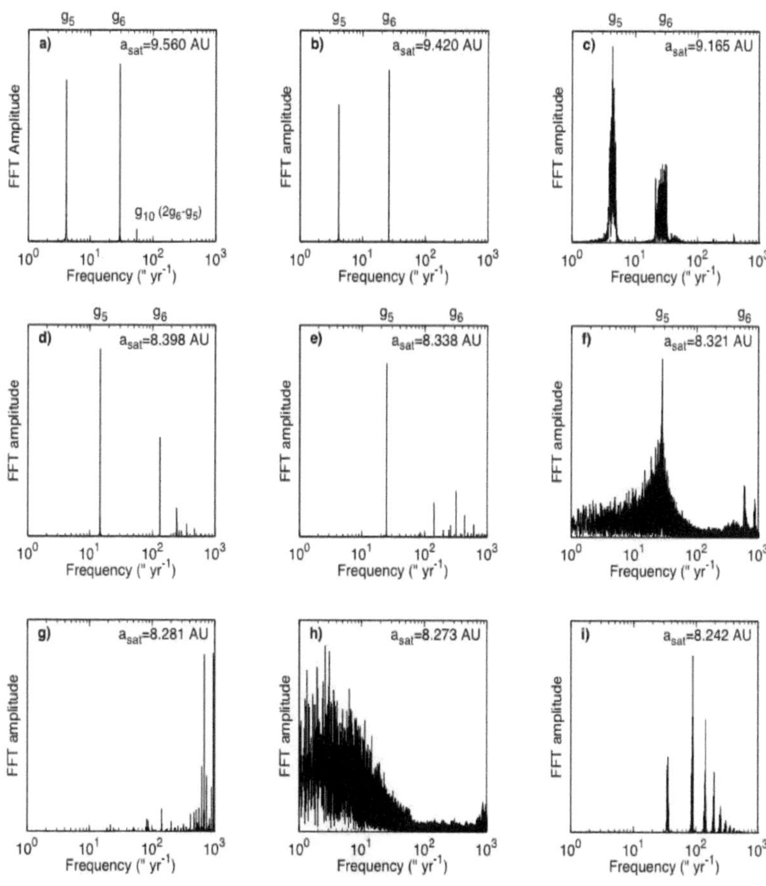

Figure 4.1: Example power spectra of Saturn's $e\{\sin\varpi, \cos\varpi\}$ time history. In each example above, Jupiter's initial semimajor axis was fixed at 5.2 AU and Saturn's semimajor axis was varied.

Table 4.1: Secular frequencies g_5 and g_6 for a migrating Saturn, with Jupiter fixed.

$a_{sat,0}$ (AU)	$\bar{a}_{jup}$ (AU)	$\bar{a}_{sat}$ (AU)	g_5 ($''$ yr^{-1})	g_6 ($''$ yr^{-1})
7.3378125	5.1978202	7.3847125	9.838	277.846
7.3500000	5.1981709	7.3946264	9.764	244.968
7.3656250	5.1985757	7.4076312	9.660	216.829
7.3900000	5.1990948	7.4286564	9.512	191.714
7.3934375	5.1991490	7.4317479	9.492	189.400
7.4000000	5.1992501	7.4376662	9.467	185.592
7.4028125	5.1992927	7.4402074	9.443	184.180
7.4156250	5.1995062	7.4516457	9.413	179.967
7.4212500	5.1994083	7.4579498	9.364	179.048
7.4284375	5.1993523	7.4655983	9.354	177.100
7.4412500	5.1994832	7.4775951	9.314	174.638
7.4490625	5.1995175	7.4852227	9.304	174.539
7.4500000	5.1995201	7.4861483	9.307	174.592
7.4540625	5.1995286	7.4901790	9.304	174.954
7.4668750	5.1994905	7.5033422	9.314	178.632
7.4768750	5.1993828	7.5141663	9.354	186.236
7.4796875	5.1993211	7.5174250	9.383	190.033
7.4925000	5.1792882	7.6748349	9.265	176.022
7.5000000	5.1879766	7.6195811	9.270	210.645
7.5046875	5.1838290	7.6546179	9.314	232.234
7.5053125	5.1824265	7.6652988	9.146	180.837
7.5181250	5.1867414	7.6472764	9.423	205.873
7.5309375	5.1929223	7.6155142	9.423	175.795
7.5325000	5.1864155	7.6644183	9.552	208.998
7.5437500	5.1957835	7.6079225	8.424	185.139

Continued on next page

Table 4.1 – continued from previous page

$a_{sat,0}$ (AU)	$\bar{a}_{jup}$ (AU)	$\bar{a}_{sat}$ (AU)	g_5 ($''$ yr^{-1})	g_6 ($''$ yr^{-1})
7.5500000	5.1851589	7.6918132	9.134	160.082
7.5565625	5.1971782	7.6110798	9.136	162.070
7.5603125	5.2013665	7.5840692	9.522	155.584
7.5693750	5.2009146	7.5963695	9.126	138.093
7.5821875	5.2007172	7.6106445	8.978	132.012
7.5881250	5.2006625	7.6169995	8.948	130.756
7.5950000	5.2006216	7.6241986	8.929	129.737
7.6000000	5.2005918	7.6294364	8.911	129.195
7.6078125	5.2005446	7.6376252	8.909	128.600
7.6159375	5.2008019	7.6439443	8.889	127.394
7.6206250	5.2005279	7.6506215	8.909	127.730
7.6334375	5.2004708	7.6639086	8.929	127.335
7.6437500	5.2004230	7.6746214	8.958	127.463
7.6462500	5.2004090	7.6772360	8.978	127.612
7.6500000	5.2003824	7.6811992	8.998	128.009
7.6590625	5.2002492	7.6912807	9.136	131.577
7.6715625	5.1999095	7.7063670	8.770	128.937
7.6718750	5.1999325	7.7064960	7.841	97.810
7.6846875	5.2013650	7.7088438	8.840	102.793
7.6975000	5.2004818	7.7282200	9.047	123.014
7.6993750	5.2004667	7.7302161	9.037	122.757
7.7000000	5.2004622	7.7308773	9.035	122.682
7.7103125	5.2004064	7.7416562	9.047	121.926
7.7231250	5.2003602	7.7548802	9.097	121.422
7.7271875	5.2003835	7.7587937	9.107	121.244
7.7359375	5.2003197	7.7680604	9.156	121.086
7.7487500	5.2002869	7.7811927	9.225	120.888

Continued on next page

Table 4.1 – continued from previous page

$a_{sat,0}$ (AU)	$\bar{a}_{jup}$ (AU)	$\bar{a}_{sat}$ (AU)	g_5 ($''$ yr^{-1})	g_6 ($''$ yr^{-1})
7.7500000	5.2002847	7.7824679	9.233	120.890
7.7550000	5.2002780	7.7875460	9.265	120.898
7.7615625	5.2002851	7.7940925	9.324	121.293
7.7743750	5.2001144	7.8082650	9.413	121.046
7.7828125	5.2001071	7.8168097	9.443	120.700
7.7871875	5.2000966	7.8212914	9.472	120.690
7.8000000	5.2000542	7.8345055	9.579	120.852
7.8106250	5.2000120	7.8455192	9.680	121.105
7.8384375	5.1998802	7.8745259	9.997	122.381
7.8500000	5.1998185	7.8866447	10.147	123.201
7.8662500	5.1997208	7.9037676	10.392	124.724
7.8940625	5.1995214	7.9333474	10.916	128.670
7.9000000	5.1994724	7.9397170	11.050	129.788
7.9218750	5.1992673	7.9633938	11.598	135.047
7.9496875	5.1989285	7.9941695	12.498	145.538
7.9500000	5.1989252	7.9945432	12.520	145.695
7.9775000	5.1984446	8.0261765	13.744	163.999
8.0000000	5.1978399	8.0538974	15.141	192.118
8.0053125	5.1976452	8.0608907	15.544	202.314
8.0081250	5.1975286	8.0647073	15.771	208.622
8.0162500	5.1971242	8.0763196	16.522	232.046
8.0243750	5.1965381	8.0894873	17.452	269.254
8.0325000	5.1954451	8.1070201	18.836	348.237
8.0331250	5.1953001	8.1088763	19.004	359.490
8.0406250	5.1800061	8.2452308	14.337	—
8.0487500	5.1809703	8.2449174	20.280	—
8.0500000	5.1814947	8.2415551	35.336	—

Continued on next page

Table 4.1 – continued from previous page

$a_{sat,0}$ (AU)	$\bar{a}_{jup}$ (AU)	$\bar{a}_{sat}$ (AU)	g_5 ($''$ yr^{-1})	g_6 ($''$ yr^{-1})
8.0568750	5.1822251	8.2422541	37.396	—
8.0609375	5.1826513	8.2427116	37.742	—
8.0650000	5.1830748	8.2431856	37.405	—
8.0731250	5.1830784	8.2513205	24.067	—
8.0812500	5.1834888	8.2558731	27.112	—
8.0887500	5.1837568	8.2609557	29.485	—
8.0893750	5.1837194	8.2619802	29.258	—
8.0975000	5.1856789	8.2534225	26.499	—
8.1000000	5.1858706	8.2542577	27.278	—
8.1056250	5.1863328	8.2558951	28.744	—
8.1137500	5.1870268	8.2580485	31.166	—
8.1165625	5.1872760	8.2587159	32.333	—
8.1218750	5.1852281	8.2814409	27.429	—
8.1300000	5.1863833	8.2797257	22.603	—
8.1381250	5.1877928	8.2759569	—	—
8.1443750	5.1885999	8.2754593	—	—
8.1462500	5.1888193	8.2755221	—	—
8.1500000	5.1894846	8.2736543	—	—
8.1543750	5.1900137	8.2735927	—	—
8.1625000	5.1910679	8.2728818	—	—
8.1706250	5.1919233	8.2739240	—	—
8.1721875	5.1921294	8.2737254	—	—
8.1787500	5.1929074	8.2738021	—	—
8.1868750	5.1937412	8.2749924	—	—
8.2000000	5.1951943	8.2759751	—	—
8.2500000	5.2002363	8.2838192	—	—
8.3000000	5.2048105	8.2952833	—	—

Continued on next page

Table 4.1 – continued from previous page

$a_{sat,0}$ (AU)	$\bar{a}_{jup}$ (AU)	$\bar{a}_{sat}$ (AU)	g_5 ($''$ yr^{-1})	g_6 ($''$ yr^{-1})
8.3190625	5.2061221	8.3033413	—	—
8.3281250	5.2073221	8.3020797	—	—
8.3371875	5.2077003	8.3080275	—	—
8.3462500	5.2080905	8.3139110	—	—
8.3500000	5.2082785	8.3161005	26.066	—
8.3525000	5.2084737	8.3169855	26.568	907.687
8.3550000	5.2085880	8.3185682	32.471	893.024
8.3553125	5.2085384	8.3192391	24.512	614.752
8.3575000	5.2086416	8.3206149	29.881	886.755
8.3600000	5.2088690	8.3212616	27.735	589.005
8.3625000	5.2071796	8.3379462	24.532	319.148
8.3643750	5.2073070	8.3387764	23.305	320.206
8.3650000	5.2091419	8.3239903	24.680	373.461
8.3675000	5.2071529	8.3432345	21.486	294.132
8.3734375	5.2069631	8.3508116	18.896	257.715
8.3825000	5.2066152	8.3628696	16.987	211.410
8.3915625	5.2062721	8.3748701	15.989	176.774
8.4000000	5.2059766	8.3858374	15.338	152.530
8.4006250	5.2059559	8.3866401	15.296	150.976
8.4096875	5.2056752	8.3981003	14.733	131.626
8.4187500	5.2054288	8.4092740	14.229	116.834
8.4278125	5.2052127	8.4201885	13.754	105.305
8.4368750	5.2050224	8.4308839	13.319	96.139
8.4459375	5.2048534	8.4413993	12.904	88.733
8.4500000	5.2047839	8.4460599	12.730	85.875
8.4550000	5.2047029	8.4517588	12.518	82.662
8.4640625	5.2045675	8.4619898	12.142	77.609

Continued on next page

Table 4.1 – continued from previous page

$a_{sat,0}$ (AU)	$\bar{a}_{jup}$ (AU)	$\bar{a}_{sat}$ (AU)	g_5 ($''$ yr^{-1})	g_6 ($''$ yr^{-1})
8.4731250	5.2044450	8.4721115	11.796	73.357
8.4821875	5.2043334	8.4821409	11.460	69.738
8.4912500	5.2042315	8.4920890	11.144	66.634
8.5000000	5.2041413	8.5016247	10.852	64.035
8.5003125	5.2041381	8.5019659	10.847	63.944
8.5093750	5.2040520	8.5117789	10.560	61.601
8.5184375	5.2039723	8.5215393	10.293	59.544
8.5275000	5.2038983	8.5312512	10.036	57.715
8.5365625	5.2038294	8.5409194	9.789	56.093
8.5456250	5.2037651	8.5505505	9.561	54.630
8.5500000	5.2037355	8.5551874	9.455	53.987
8.5546875	5.2037048	8.5601440	9.344	53.315
8.5637500	5.2036483	8.5697109	9.146	52.128
8.5728125	5.2035948	8.5792497	8.948	51.051
8.5818750	5.2035449	8.5887587	8.761	50.052
8.5909375	5.2034975	8.5982455	8.592	49.142
8.6000000	5.2034527	8.6077111	8.424	48.302
8.6500000	5.2032424	8.6596227	7.651	44.594
8.7000000	5.2030801	8.7111276	7.045	41.899
8.7500000	5.2029503	8.7623590	6.575	39.786
8.8000000	5.2028797	8.8130503	6.180	37.907
8.8500000	5.2027567	8.8642631	5.883	36.486
8.9000000	5.2026842	8.9150081	5.611	35.138
8.9500000	5.2026066	8.9658181	5.389	33.890
9.0000000	5.2025558	9.0163764	5.191	32.753
9.0500000	5.2025080	9.0669157	5.018	31.727
9.1000000	5.2024685	9.1173807	4.870	30.837

Continued on next page

Table 4.1 – continued from previous page

$a_{sat,0}$ (AU)	$\bar{a}_{jup}$ (AU)	$\bar{a}_{sat}$ (AU)	g_5 ($''$ yr^{-1})	g_6 ($''$ yr^{-1})
9.1125000	5.2024614	9.1299698	4.835	30.692
9.1250000	5.2024571	9.1425292	4.815	30.702
9.1375000	5.2024409	9.1552144	4.865	31.947
9.1500000	5.2026839	9.1652212	4.783	30.405
9.1625000	5.2023813	9.1808669	4.687	29.644
9.1750000	5.2023811	9.1933868	4.637	29.258
9.1875000	5.2023768	9.2059505	4.608	29.021
9.2000000	5.2023705	9.2185351	4.585	28.810
9.2500000	5.2023401	9.2689323	4.462	28.019
9.3000000	5.2023082	9.3193532	4.351	27.290
9.3500000	5.2022756	9.3697889	4.252	26.623
9.4000000	5.2022401	9.4202636	4.165	26.054
9.4046875	5.2022365	9.4249992	4.153	26.005
9.4093750	5.2022327	9.4297361	4.153	25.968
9.4140625	5.2022290	9.4344735	4.140	25.918
9.4187500	5.2022251	9.4392123	4.140	25.881
9.4234375	5.2022211	9.4439527	4.128	25.844
9.4281250	5.2022170	9.4486941	4.116	25.807
9.4328125	5.2022127	9.4534371	4.116	25.782
9.4375000	5.2022083	9.4581819	4.103	25.745
9.4421875	5.2022038	9.4629285	4.103	25.720
9.4468750	5.2021990	9.4676774	4.091	25.696
9.4500000	5.2021957	9.4708452	4.091	25.683
9.4515625	5.2021941	9.4724288	4.091	25.683
9.4562500	5.2021888	9.4771826	4.079	25.671
9.4609375	5.2021833	9.4819404	4.066	25.659
9.4656250	5.2021773	9.4867347	4.066	25.659

Continued on next page

Table 4.1 – continued from previous page

$a_{sat,0}$ (AU)	$\bar{a}_{jup}$ (AU)	$\bar{a}_{sat}$ (AU)	g_5 ($''$ yr^{-1})	g_6 ($''$ yr^{-1})
9.4703125	5.2021713	9.4914644	4.066	25.671
9.4750000	5.2021647	9.4962356	4.054	25.683
9.4796875	5.2021576	9.5010123	4.054	25.708
9.4843750	5.2021499	9.5057956	4.042	25.745
9.4890625	5.2021414	9.5105869	4.042	25.795
9.4937500	5.2021321	9.5153884	4.042	25.869
9.4984375	5.2021218	9.5202016	4.029	25.955
9.5000000	5.2021180	9.5218089	4.029	25.992
9.5031250	5.2021101	9.5250290	4.029	26.079
9.5078125	5.2020968	9.5298757	4.029	26.227
9.5125000	5.2020813	9.5347465	4.029	26.425
9.5171875	5.2020631	9.5396493	4.029	26.697
9.5218750	5.2020409	9.5445962	4.029	27.055
9.5265625	5.2020133	9.5496056	4.042	27.537
9.5312500	5.2019772	9.5547105	4.054	28.242
9.5359375	5.2019270	9.5599748	4.066	29.305
9.5406250	5.2018492	9.5655524	4.091	31.146
9.5453125	5.2016924	9.5720250	4.153	35.373
9.5500000	5.2002731	9.5928086	4.153	36.251
9.5546875	5.2006251	9.5935278	5.426	—
9.5593750	5.2009735	9.5942701	3.967	—
9.5640625	5.2013193	9.5950439	4.548	—
9.5687500	5.2015762	9.5968308	4.153	—
9.5734375	5.2019044	9.5978075	4.252	—
9.5781250	5.2022616	9.5984534	4.264	—
9.5828125	5.2026565	9.5986711	4.178	—
9.5875000	5.2031047	9.5982851	4.005	—

Continued on next page

Table 4.1 – continued from previous page

$a_{sat,0}$ (AU)	$\bar{a}_{jup}$ (AU)	$\bar{a}_{sat}$ (AU)	g_5 ($''$ yr^{-1})	g_6 ($''$ yr^{-1})
9.5921875	5.2034748	9.5987863	3.930	—
9.5968750	5.2027443	9.6117609	4.029	25.733
9.6000000	5.2026335	9.6161483	3.943	23.978
9.6015625	5.2025969	9.6181287	3.918	23.483
9.6062500	5.2025210	9.6236858	3.856	22.593
9.6109375	5.2024720	9.6289377	3.819	22.124
9.6156250	5.2024369	9.6340321	3.794	21.852
9.6203125	5.2024101	9.6390322	3.782	21.691
9.6250000	5.2023888	9.6439707	3.770	21.568
9.6296875	5.2023712	9.6488661	3.757	21.481
9.6343750	5.2023565	9.6537303	3.745	21.419
9.6390625	5.2023438	9.6585707	3.745	21.370
9.6437500	5.2023327	9.6633930	3.733	21.320
9.6484375	5.2023229	9.6682006	3.720	21.283
9.6500000	5.2023199	9.6698004	3.720	21.271
9.6531250	5.2023142	9.6729965	3.720	21.246
9.6578125	5.2023063	9.6777828	3.708	21.209
9.6625000	5.2022991	9.6825611	3.708	21.172
9.6671875	5.2022926	9.6873325	3.696	21.135
9.6718750	5.2022865	9.6920982	3.696	21.098
9.6765625	5.2022809	9.6968589	3.683	21.061
9.6812500	5.2022756	9.7016154	3.683	21.036
9.6859375	5.2022707	9.7063681	3.671	20.999
9.6906250	5.2022661	9.7111175	3.671	20.962
9.6953125	5.2022617	9.7158641	3.658	20.925
9.7000000	5.2022576	9.7206080	3.658	20.888
9.7500000	5.2022231	9.7711041	3.597	20.468

Continued on next page

Table 4.1 – continued from previous page

$a_{sat,0}$ (AU)	$\bar{a}_{jup}$ (AU)	$\bar{a}_{sat}$ (AU)	g_5 ($''$ yr^{-1})	g_6 ($''$ yr^{-1})
9.8000000	5.2021982	9.8214942	3.535	20.035
9.8500000	5.2021780	9.8718345	3.473	19.590
9.9000000	5.2021605	9.9221474	3.424	19.145
9.9500000	5.2021445	9.9724467	3.362	18.712
10.0000000	5.2021381	10.0226309	3.312	18.317
10.0500000	5.2021179	10.0729896	3.263	17.884
10.1000000	5.2021046	10.1232661	3.226	17.501
10.1500000	5.2020919	10.1735395	3.176	17.130
10.2000000	5.2020672	10.2239703	3.139	16.772
10.2500000	5.2020666	10.2740944	3.102	16.451
10.3000000	5.2020536	10.3243825	3.078	16.154
10.3500000	5.2020398	10.3746846	3.053	15.870
10.4000000	5.2020247	10.4250077	3.028	15.635
10.4500000	5.2020075	10.4753634	3.028	15.450

Table 4.2: Secular frequencies g_5 and g_6 for a migrating Jupiter and Saturn.

$a_{jup,0}$ (AU)	$a_{sat,0}$ (AU)	$\bar{a}_{jup}$ (AU)	$\bar{a}_{sat}$ (AU)	g_5 ($''$ yr^{-1})	g_6 ($''$ yr^{-1})
5.2019111	9.5331445	5.2019591	9.5568133	4.054	28.613
5.2050857	9.5204461	5.2052481	9.5428001	4.029	26.511
5.2082603	9.5077476	5.2084764	9.5294747	4.029	25.856
5.2114349	9.4950492	5.2116850	9.5163769	4.042	25.622
5.2146095	9.4823508	5.2148844	9.5033796	4.054	25.572
5.2177841	9.4696524	5.2180790	9.4904393	4.079	25.609

Continued on next page

Table 4.2 – continued from previous page

$a_{jup,0}$ (AU)	$a_{sat,0}$ (AU)	$\bar{a}_{jup}$ (AU)	$\bar{a}_{sat}$ (AU)	g_5 ($''$ yr^{-1})	g_6 ($''$ yr^{-1})
5.2209587	9.4569540	5.2212707	9.4775329	4.103	25.708
5.2241333	9.4442556	5.2244605	9.4646478	4.128	25.844
5.2273079	9.4315572	5.2276492	9.4517772	4.165	26.005
5.2304825	9.4188588	5.2308371	9.4389161	4.190	26.190
5.2336571	9.4061603	5.2340244	9.4260618	4.215	26.388
5.2368317	9.3934619	5.2372115	9.4132122	4.252	26.598
5.2400063	9.3807635	5.2403981	9.4003678	4.289	26.820
5.2431809	9.3680651	5.2435849	9.3875231	4.314	27.043
5.2463555	9.3553667	5.2467715	9.3746809	4.351	27.278
5.2495301	9.3426683	5.2499580	9.3618403	4.388	27.525
5.2527047	9.3299699	5.2531445	9.3490016	4.425	27.772
5.2558793	9.3172715	5.2563308	9.3361662	4.462	28.032
5.2590539	9.3045730	5.2595165	9.3233375	4.499	28.291
5.2622285	9.2918746	5.2627006	9.3105261	4.536	28.563
5.2654031	9.2791762	5.2658797	9.2977685	4.585	28.922
5.2685777	9.2664778	5.2692937	9.2825854	4.585	30.541
5.2717523	9.2537794	5.2723044	9.2715673	4.721	30.306
5.2749269	9.2410810	5.2754847	9.2588005	4.734	30.059
5.2781015	9.2283826	5.2786704	9.2459780	4.771	30.244
5.2812762	9.2156842	5.2818589	9.2331288	4.820	30.516
5.2844508	9.2029857	5.2850490	9.2202645	4.870	30.813
5.2876254	9.1902873	5.2882401	9.2073906	4.919	31.122
5.2908000	9.1775889	5.2914314	9.1945169	4.981	31.455
5.2939746	9.1648905	5.2946250	9.1816200	5.030	31.789
5.2971492	9.1521921	5.2978187	9.1687252	5.092	32.135
5.3003238	9.1394937	5.3010127	9.1558280	5.154	32.506
5.3034984	9.1267953	5.3042058	9.1429414	5.228	32.877

Continued on next page

Table 4.2 – continued from previous page

$a_{jup,0}$ (AU)	$a_{sat,0}$ (AU)	$\bar{a}_{jup}$ (AU)	$\bar{a}_{sat}$ (AU)	g_5 ($''$ yr^{-1})	g_6 ($''$ yr^{-1})
5.3066730	9.1140968	5.3073824	9.1302179	5.315	33.445
5.3098476	9.1013984	5.3106195	9.1169011	5.376	33.705
5.3130222	9.0887000	5.3138135	9.1040104	5.451	34.076
5.3161968	9.0760016	5.3170131	9.0910674	5.525	34.496
5.3193714	9.0633032	5.3202149	9.0781039	5.611	34.928
5.3225460	9.0506048	5.3234188	9.0651225	5.710	35.386
5.3257206	9.0379064	5.3266247	9.0521234	5.809	35.855
5.3288952	9.0252080	5.3298335	9.0390999	5.908	36.350
5.3320698	9.0125095	5.3330385	9.0261139	6.019	36.881
5.3352444	8.9998111	5.3362531	9.0130396	6.143	37.425
5.3384190	8.9871127	5.3394682	8.9999627	6.266	38.006
5.3415936	8.9744143	5.3426860	8.9868635	6.402	38.624
5.3447682	8.9617159	5.3459069	8.9737379	6.551	39.279
5.3479428	8.9490175	5.3491310	8.9605864	6.711	39.996
5.3511174	8.9363191	5.3523594	8.9473977	6.897	40.762
5.3542920	8.9236207	5.3555921	8.9341723	7.082	41.603
5.3574666	8.9109222	5.3588299	8.9209047	7.292	42.542
5.3606412	8.8982238	5.3620735	8.9075881	7.515	43.580
5.3638158	8.8855254	5.3653236	8.8942161	7.762	44.754
5.3669904	8.8728270	5.3685814	8.8807794	8.046	46.101
5.3701650	8.8601286	5.3718480	8.8672674	8.343	47.659
5.3733396	8.8474302	5.3751253	8.8536665	8.676	49.488
5.3765142	8.8347318	5.3784150	8.8399597	9.060	51.676
5.3796889	8.8220334	5.3817200	8.8261242	9.467	54.345
5.3828635	8.8093349	5.3850433	8.8121334	9.925	57.670
5.3860381	8.7966365	5.3883899	8.7979462	10.432	61.897
5.3892127	8.7839381	5.3917660	8.7835104	10.988	67.409

Continued on next page

Table 4.2 – continued from previous page

$a_{jup,0}$ (AU)	$a_{sat,0}$ (AU)	$\bar{a}_{jup}$ (AU)	$\bar{a}_{sat}$ (AU)	g_5 ($''$ yr^{-1})	g_6 ($''$ yr^{-1})
5.3923873	8.7712397	5.3951804	8.7687500	11.606	74.850
5.3955619	8.7585413	5.3986461	8.7535558	12.298	85.244
5.3987365	8.7458429	5.4021823	8.7377655	13.052	100.484
5.4019111	8.7331445	5.4058161	8.7211472	13.917	124.165

Fig. 4.1a shows a result of a simulation whose initial conditions were most like the present solar system. The well known g_5 and g_6 frequencies are easily identified as the two strongest peaks in the spectrum. The labels in Fig. 4.1 were placed along the x-axis at the frequencies given by **Brouwer and van Woerkom (1950)**. One additional peak is visible at the location of the frequency $g_{10} = 2g_6 - g_5$, which arises due to the proximity of Jupiter and Saturn to the 5:2 resonance. Fig. 4.1b shows an example spectrum with Saturn's semimajor axis displaced inwards by 0.14 AU, and the g_{10} peak is no longer visible. Jupiter and Saturn are within the 7:3 resonance in Fig. 4.1c, and while the g_5 and g_6 frequencies are still identifiable, the spectrum shows the characteristic broadband noise of orbital chaos. One notable trend observed in Figs. 4.1c–f is that as Saturn and Jupiter approach the 2:1 resonance, the g_6 frequency becomes less dominant compared with the g_5 in the spectrum of Saturn's (h, k).

The spectral character of Saturn's orbital history within the 2:1 resonance with Jupiter can be quite complex. Figs. 4.1g–i show examples of spectra for orbits with slightly different initial conditions, but still very close to the 2:1 resonance. Orbits can be chaotic, as in Fig. 4.1h, or regular, such as in Figs. 4.1g and i. But even the regular orbits have lost the characteristic g_5 and g_6 frequencies.

In all 255 simulations, when the g_5 and g_6 frequencies were identifiable their value as a function of Saturn's mean semimajor axis was obtained and the results shown in Fig. 4.2a. An additional set of 64 simulations were carried out in which both

Jupiter and Saturn's semimajor axes were varied from 5.2–5.4 AU and 9.6–8.7 AU, respectively, and the resulting g_5 and g_6 frequencies were identified and plotted in Fig. 4.2b. This migration path is the same as that used in **Minton and Malhotra** (2009), and the g_5 and g_6 frequency values are plotted in Fig. 4.2b.

4.3 Analytical calculations of the g_5 and g_6 frequencies

The well known first order Laplace-Lagrange secular theory provides a simple and direct way of calculating the secular frequencies. In this theory, only the terms of the disturbing function are retained of second order in eccentricity and first order in mass (**Murray and Dermott, 1999**). In the planar two-planet case, the secular perturbations of planet $j =$, where $j = 5$ is Jupiter and $j = 6$ is Saturn, are described by the first order secular terms of the disturbing function:

$$R_j = \frac{n_j}{a_j^2} \left[\frac{1}{2} A_{jj} e_j^2 + A_{jk} e_5 e_6 \cos(\varpi_1 - \varpi_2) \right], \tag{4.2}$$

where n is the mean motion, and $\mathbf{A}$ is a matrix with elements

$$A_{jj} = +n_j \frac{1}{4} \frac{m_k}{M_\odot + m_j} \alpha_{jk} \bar{\alpha}_{jk} b_{3/2}^{(1)}(\alpha_{jk}), \tag{4.3}$$

$$A_{jk} = -n_j \frac{1}{4} \frac{m_k}{M_\odot + m_j} \alpha_{jk} \bar{\alpha}_{jk} b_{3/2}^{(2)}(\alpha_{jk}), \tag{4.4}$$

for $j = 5, 6$, $k = 6, 5$, and $j \neq k$; $\alpha_{jk} = \min\{a_j/a_k, a_j/a_k\}$, and

$$\bar{\alpha}_{jk} = \begin{cases} 1 & : a_j > a_k \\ a_j/a_k & : a_j < a_k, \end{cases} \tag{4.5}$$

and $b_{3/2}^{(1)}$ and $b_{3/2}^{(2)}$ are Laplace coefficients. The secular motion of the system can be described as a set of linear differential equations in terms of the $\{h, k\}$ vectors given in Eq. (4.1):

$$\dot{h}_j = +\sum_{p=5}^{6} A_{pj} k_j, \quad \dot{k}_j = -\sum_{p=5}^{6} A_{pj} h_j. \tag{4.6}$$

The secular frequencies g_5 and g_6 are the eigenvalues of the matrix $\mathbf{A}$, therefore only depend on the relative masses of Jupiter and Saturn to the Sun, and the semimajor axes of Jupiter and Saturn. Using the current semimajor axes of Jupiter and Saturn the theory gives frequency values $g_5 = 3.7'' \, \mathrm{yr}^{-1}$ and $g_6 = 22.3'' \, \mathrm{yr}^{-1}$, which are lower than the more accurate values given by **Brouwer and van Woerkom (1950)** by 14% and 20%, respectively (**Laskar, 1988**). **Brouwer and van Woerkom (1950)** achieved their more accurate solution by incorporating higher order terms in the disturbing function involving $2\lambda_5 - 5\lambda_6$, which arise due to Jupiter and Saturn's proximity to the 5:2 resonance (the so-called "Great Inequality"). However, as Fig. 4.1b and Fig. 4.2a illustrate, the effect of the 5:2 resonance is only important over a very narrow range in Saturn's semimajor axis.

The dash-dot line in Fig. 4.2 shows the calculation of the g_5 and g_6 frequencies as a function of Saturn's semimajor axis (with Jupiter fixed at 5.2 AU) using first order Laplace-Lagrange secular theory. The theory is adequate in the vicinity of Saturn's current position (but away from the 5:2 resonance), but does a poor job near the 2:1 resonance. **Malhotra et al. (1989)** developed corrections for the first order Laplace-Lagrange theory to account for the perturbations from $n + 1 : n$ resonances in the context of the Uranian satellite system. The corrections for the near 2:1 resonance between Jupiter and Saturn are additional terms that are added to elements of the **A** matrix:

$$\tilde{A}_{jj} = A_{jj} + \varepsilon n_5 \tfrac{m_6}{M_\odot} \alpha C_1 C_2 \tag{4.7}$$

$$\tilde{A}_{jk} = A_{jk} + \varepsilon n_6 \tfrac{m_5}{M_\odot} C_1 C_2, \tag{4.8}$$

where the coefficients C_1 and C_2 are

$$C_1(\alpha) = -\tfrac{1}{2} \left(4 + \alpha \tfrac{d}{d\alpha}\right) b_{1/2}^{(2)}(\alpha) \tag{4.9}$$

$$C_2(\alpha) = +\tfrac{1}{2} \left(3 + \alpha \tfrac{d}{d\alpha}\right) b_{1/2}^{(1)}(\alpha), \tag{4.10}$$

and

$$\varepsilon = \frac{3}{2} \frac{\left[\tfrac{m_6}{M_\odot} \left(F_{11} - 2F_{21}\right) \alpha^{-2} + 2 \tfrac{m_5}{M_\odot} \left(2F_{22} - F_{12}\right) \right]}{\left(\dot{\lambda}_5 - 2\dot{\lambda}_6\right)^2 / n_6^2} \tag{4.11}$$

where

$$F_{11} = 1 + \frac{m_6}{M_\odot}\alpha\left(\frac{1}{3}\alpha\frac{d}{d\alpha} + \frac{2}{3}\alpha^2\frac{d^2}{d\alpha^2}\right)b_{1/2}^{(0)} \tag{4.12}$$

$$F_{12} = -\frac{2}{3}\frac{m_6}{M_\odot}\alpha^{-1/2}\left(2\alpha\frac{d}{d\alpha} + \alpha^2\frac{d^2}{d\alpha^2}\right)b_{1/2}^{(0)} \tag{4.13}$$

$$F_{21} = -\frac{2}{3}\frac{m_5}{M_\odot}\alpha^{3/2}\left(2\alpha\frac{d}{d\alpha} + \alpha^2\frac{d^2}{d\alpha^2}\right)b_{1/2}^{(0)} \tag{4.14}$$

$$F_{22} = 1 + \frac{m_5}{M_\odot}\left(1 + \frac{7}{3}\alpha\frac{d}{d\alpha} + \frac{2}{3}\alpha^2\frac{d^2}{d\alpha^2}\right)b_{1/2}^{(0)}, \tag{4.15}$$

and the $\dot{\lambda}_i$ are

$$\dot{\lambda}_5 = n_5\left[1 - \frac{m_6}{M_\odot}\alpha^2\frac{d}{d\alpha}b_{1/2}^{(0)}\right] \tag{4.16}$$

$$\dot{\lambda}_6 = n_6\left[1 + \frac{m_5}{M_\odot}\left(1 + \alpha\frac{d}{d\alpha}\right)b_{1/2}^{(0)}\right]. \tag{4.17}$$

These corrections are first order in mass, so the correction should be valid for the giant planets where the Sun-Jupiter mass ratio is $\mathcal{O}(10^{-3})$. Importantly, these corrections are zeroth order in eccentricity, therefore the nearby 2:1 frequencies (and any other first order resonance) will affect the secular frequencies even if Saturn and Jupiter were on nearly circular orbits, as in some models of the pre-migration state of the giant planets (Tsiganis et al., 2005). Higher order resonances, such as the third order 5:2, will have eccentricity dependent effects on the secular frequencies that are $\mathcal{O}(e^{j-1})$, where j is the order of the resonance.

The values of the g_5 and g_6 frequencies using the near 2:1 mean motion resonance corrections are found from the eigenvalues of the $\tilde{\mathbf{A}}$ matrix. The secular frequencies as a function of Saturn's semimajor axis are plotted as the dashed line in Fig. 4.2. The corrected values match those found from the Fourier spectral analysis of the numerical integrations for most of the range considered, with the exception of near higher order mean motion resonances. Therefore the first order Laplace-Lagrange secular theory with the corrections to near 2:1 mean motion resonance of Malhotra et al. (1989) is adequate to describe the changes in the secular frequencies during giant planet migration during of most of the giant planet migration that is relevant for considering the effects of the asteroid belt.

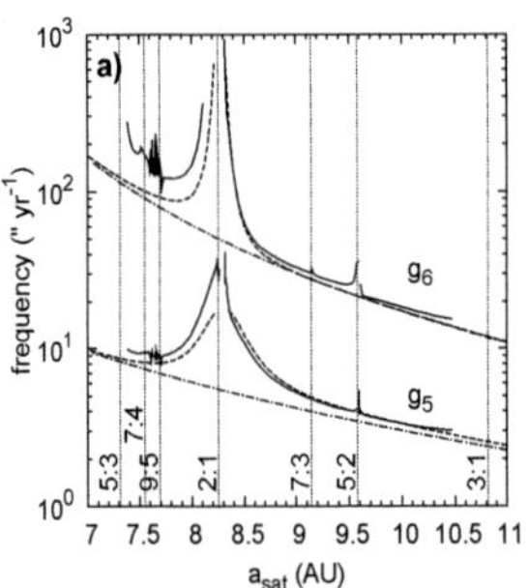
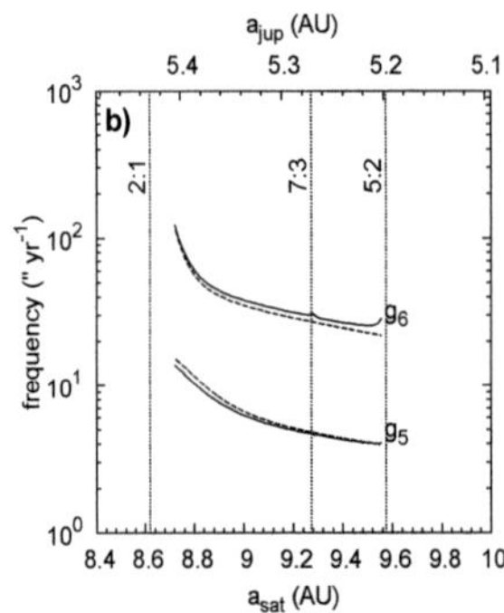

Figure 4.2: The g_5 and g_6 frequencies as a function of Jupiter and Saturn's semimajor axis. The locations of Jupiter/Saturn MMRs which have an effect on the location of the ν_6 in the range explored numerically are shown. **a)** The g_5 and g_6 frequencies as a function of Saturn's semimajor axis for Jupiter fixed at 5.2 AU. The solid lines were obtained by identifying the secular frequencies from the Fourier spectra of 255 ~ 100 My N-body simulations of Jupiter and Saturn, where a_{sat} was varied from 8.32–10.5 AU. The dash-dot lines are the secular frequencies obtained using first order Laplace-Lagrange secular theory. The dashed lines are the secular frequencies obtained using the first order Laplace-Lagrange secular theory with corrections to the near 2:1 mean motion resonance between Jupiter and Saturn using the method of Malhotra et al. (1989). **b)** The solid lines are obtained in a similar way as the points in **(a)**, but in this case Jupiter's semimajor axis was varied from 5.2–5.4 AU while Saturn's semimajor axis was varied from 9.56–8.7 AU. The dashed lines are the secular frequencies obtained using the first order Laplace-Lagrange secular theory with corrections to the near 2:1 mean motion resonance between Jupiter and Saturn using the method of Malhotra et al. (1989).

4.4 The zero-inclination location of the ν_6 resonance

The secular variation in eccentricity of an asteroid that is perturbed by the two giant planets in my simplified solar system is approximately described by a Hamiltonian which is analogous to a resonantly forced oscillator with a natural frequency g_0 and a forcing frequency $g_i \approx g_0$, and ignoring any nonlinear terms:

$$\mathcal{H} = -g_0 J + \sigma\sqrt{2J}\cos\phi, \tag{4.18}$$

where ϕ is the resonance angle, which is approximately the measures the asteroid's longitude of perihelion relative to the longitude of perihelion of the planet whose

g_i is the dominant mode. In the current solar system, g_6 is the dominant mode in Saturn's precession rate, but as Fig. 4.1 suggests, some configurations lead to other modes being dominant. The variable J is the canonically conjugate generalized momentum which is related to the asteroid's orbital semimajor axis a and eccentricity e, $J = \sqrt{a}\left(1 - \sqrt{1-e^2}\right)$. Because a is unchanged by the secular perturbation, the dynamical changes in J due to the secular perturbation reflect changes in the asteroid's eccentricity e. If, for simplicity, I neglect the influence of all planets except for Jupiter and Saturn, and only consider the effect of the g_6 frequency, then the coefficient σ is defined as:

$$\sigma = \frac{1}{4}\frac{1}{a^{5/4}}\sum_i \alpha_i^2 b_{3/2}^{(2)}(\alpha_i) m_i E_i^{(6)}, \tag{4.19}$$

where $\alpha_j = a_0/a_j$ and a_0 is the asteroid semimajor axis in units of AU. The coefficient σ is proportional to $E_j^{(6)}$, which is the power of the g_6 mode in the planet j, and is related to the eccentricity of the giant planets.

Using Poincaré variables, $(x, y) = \sqrt{2J}(\cos\phi, -\sin\phi)$, the equations of motion can be derived from this Hamiltonian. The equations of motion are:

$$\dot{x} = (g_6 - g_0)\,y \tag{4.20}$$

$$\dot{y} = x\,(g_6 - g_0) - \sigma \tag{4.21}$$

These equations have the following solution near resonance:

$$\{x(t), y(t)\} = \frac{\sigma}{g_0 - g_6}\{\cos(g_6 t + \beta_0), -\sin(g_6 t + \beta_0)\}, \tag{4.22}$$

The ν_6 secular resonance occurs when $g_0 = g_6$, in which case the solution is that of a resonantly forced oscillator whose amplitude (the asteroid's eccentricity) grows without bounds.

By using the values of the g_6 frequency as a function of Saturn and Jupiter's semimajor axis, shown in Fig. 4.2, I can obtain the location in semimajor axis of the ν_6 resonance. One complicating factor is that when taking into account higher order terms in mass and eccentricity, there is a coupling between the e–ϖ modes and the i–Ω modes, so that instead of being located at a single semimajor axis, the location of the ν_6 resonance forms surfaces in a–e–$\sin i$ space (**Williams and Faulkner, 1981**). Therefore my calculations of the position of the ν_6 are only strictly valid in the case of an asteroid with zero-inclination.

The result of these calculations relating the zero-inclination position of the ν_6 as a function of Saturn's position is shown in Fig. **4.3**. For much of the asteroid belt, the difference between the position of the ν_6 as predicted by the first order Laplace-Lagrange secular theory with near 2:1 corrections is very small — for $a_{sat} = 9.17$ AU the difference between a_{ν_6} calculated numerically vs. with the corrected first order secular theory is $\sim 5\%$. The effect of the 5:2 mean motion resonance on the value of the g_6 secular frequency (and hence the position of the ν_6 resonance location) is only apparent in a very narrow region that is currently not very populated by asteroids today.

4.5 Conclusion

I have used spectral analysis of the e-ϖ time history of Saturn from long-term numerical integrations to identify the relationship between semimajor axis locations of Jupiter and Saturn and the g_5 and g_6 secular frequencies, as well as the zero-inclination position of the important ν_6 secular resonance. The first order Laplace-Lagrange secular theory with corrections for the near 2:1 mean motion resonance between Jupiter and Saturn is adequate to describe the behavior of the secular frequencies over a large range in Saturn's semimajor axis. The exception to this is in the current solar system, where the effects of the near 5:2 resonance, the "Great Inequality," increase the secular frequencies by a substantial amount and introduce additional frequencies. When Jupiter and Saturn are within a resonance, such as the 5:2, 7:3, 2:1, and others, the e-ϖ secular frequencies broaden and the power spectra become noisy. Often in such cases it can be difficult to identify the equivalent g_5 and g_6 frequencies, and, especially near the 2:1 resonance, the g_5 mode dominates over the g_6 in Saturn's e-ϖ time history. The strength of the ν_6 resonance is related to the power of the g_6 mode, and in power spectra analysis the power of a mode

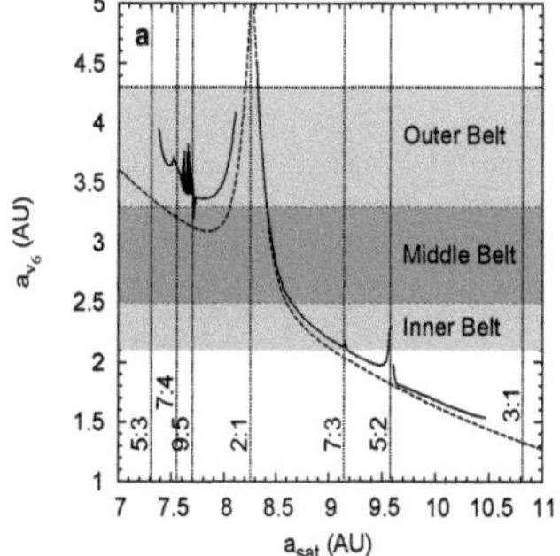

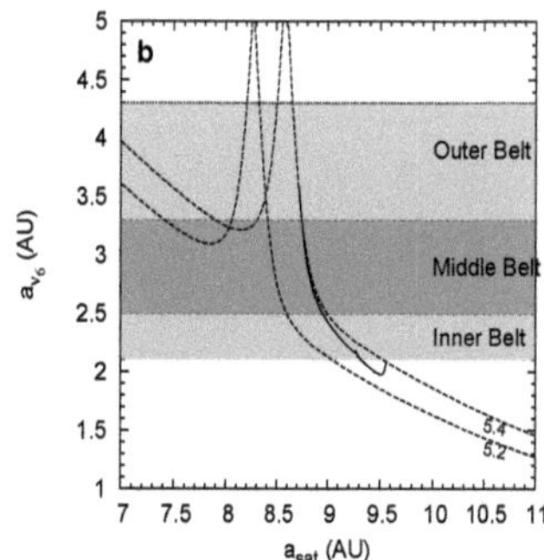

Figure 4.3: The position of the ν_6 resonance (at zero inclination) as a function of Jupiter and Saturn's semimajor axes. **a)** The dashed line shows the calculation using linear secular theory, with a correction for the effect of the near 2:1 mean motion resonance between Jupiter and Saturn (**Malhotra et al., 1989**). The solid line was obtained by using the g_6 eigenfrequencies obtained from spectral analysis of 255 asteroid from Fig. **4.2**a and then calculating the positions where $g_0 - g_6 = 0$ (for a particle at zero inclination) at each a_{sat} location. The locations of Jupiter/Saturn MMRs which have an effect on the location of the ν_6 in the range explored numerically are shown. **b)** Using the linear secular theory as in **a)**, the two solid lines show the result for two different values of Jupiter's semimajor axis, labeled in AU. The points are obtained in a similar way as the points in **(a)**, but in this case Jupiter's semimajor axis was varied from 5.2–5.4 AU while Saturn's semimajor axis was varied from 9.6–8.7 AU as in Fig. **4.2**b.

is proportional to the amplitude of the peak at that mode. Therefore, if Saturn and Jupiter were once closer to the 2:1 resonance than today, such as in Figs. **4.1**c–f where the g_5 mode dominates the power spectra, the ν_6 may not have been as powerful as today.

The particular details of the relative powers in the g_i modes in the giant planets is very dependent on choice of initial conditions. Here I have chosen initial conditions for Jupiter and Saturn in my simulations that are similar to their current configurations, changing only the initial semimajor axes of the planets and setting inclinations to zero. The details of the migration history of the giant planets, including the eccentricities of the planets and their apsidal phases, are important for determining the effect of the ν_6 resonance on the asteroid belt during migration.

CHAPTER 5

AN ANALYTICAL MODEL FOR THE SWEEPING ν_6 RESONANCE

5.1 Analytical theory of a sweeping secular resonance

I adopt a simplified model in which the main belt asteroids are perturbed only by the ν_6 resonance. I use a system of units where the mass is in solar masses, the semimajor axis is in units of AU, and the unit of time is $(1/2\pi)$y. With this system the gravitational constant, G, is unity. An asteroid's secular perturbations close to a secular resonance can be described by the following Hamiltonian function (Malhotra, 1998):

$$H_{sec} = -g_0 J + \sigma\sqrt{2J}\cos(w_p - \varpi), \tag{5.1}$$

where $w_p = g_p t + \beta_p$ describes the phase of the p-th eigenmode of the linearized eccentricity-pericenter secular theory for the Solar system planets (Murray and Dermott, 1999), g_p is the associated eigenfrequency, ϖ is the asteroid's longitude of perihelion, $J = \sqrt{a}\left(1 - \sqrt{1-e^2}\right)$ is the canonical momentum which is related to the asteroid's orbital semimajor axis a and eccentricity e; J and $-\varpi$ are the canonically conjugate pair of variables in this 1-degree-of-freedom Hamiltonian system. The coefficients g_0 and σ are given by:

$$g_0 = \frac{1}{4}\frac{1}{a^{3/2}}\sum_j \alpha_j^2 b_{3/2}^{(1)}(\alpha_j)m_j, \tag{5.2}$$

$$\sigma = \frac{1}{4}\frac{1}{a^{5/4}}\sum_j \alpha_j^2 b_{3/2}^{(2)}(\alpha_j)m_j E_j^{(p)}, \tag{5.3}$$

where j is the planet, $E_j^{(p)}$ is the amplitude of the g_p mode in the jth planet's orbit, $\alpha_j = \min\{a/a_j, a_j/a\}$, m_j is the ratio of the mass of planet j to the Sun, and $b_{3/2}^{(1)}(\alpha_j)$ and $b_{3/2}^{(2)}(\alpha_j)$ are Laplace coefficients.

It is useful to make a canonical transformation to new variables (ϕ, P) defined by the following generating function,

$$\mathcal{F}(-\varpi, P, t) = (w_p(t) - \varpi)P \tag{5.4}$$

Thus, $\phi = \partial\mathcal{F}/\partial P = (w_p(t) - \varpi)$ and $J = -\partial\mathcal{F}/\partial\varpi = P$. The new Hamiltonian function is $\tilde{H}_{sec} = H_{sec} + \partial\mathcal{F}/\partial t$,

$$\tilde{H}_{sec} = (\dot{w}_p(t) - g_0)J + \sigma\sqrt{2J}\cos\phi, \tag{5.5}$$

where I have retained J to denote the canonical momentum, since $P = J$. It is useful to make a second canonical transformation to canonical eccentric variables,

$$x = \sqrt{2J}\cos\phi, y = -\sqrt{2J}\sin\phi, \tag{5.6}$$

where x is the canonical coordinate and y is the canonically conjugate momentum. The Hamiltonian expressed in these variables is

$$\tilde{H}_{sec} = (\dot{w}_p(t) - g_0)\frac{x^2 + y^2}{2} + \sigma x. \tag{5.7}$$

During planetary migration, the secular frequency g_p is a slowly varying function of time. I approximate its rate of change, $\dot{g}_p = \lambda$, as a constant, so that

$$\dot{w}_p(t) = g_{p,0} + 2\lambda t. \tag{5.8}$$

I define $t = 0$ as the epoch of exact resonance crossing, so that $g_{p,0} = g_0$ (cf. Ward et al., 1976). Then, $\dot{w}_p(t) - g_0 = 2\lambda t$, and the equations of motion from the Hamiltonian 5.7 can be written as:

$$\dot{x} = 2y\lambda t, \tag{5.9}$$

$$\dot{y} = -2x\lambda t - \sigma. \tag{5.10}$$

These equations of motion form a system of linear, nonhomogenous differential equations, whose solution is a linear combination of a homogeneous and a particular solution. The homogeneous solution can be found by inspection, giving:

$$x_h(t) = c_1 \cos\lambda t^2 + c_2 \sin\lambda t^2, \tag{5.11}$$

$$y_h(t) = -c_1 \sin\lambda t^2 + c_2 \cos\lambda t^2, \tag{5.12}$$

where c_1 and c_2 are constant coefficients. I use the method of variation of parameters to find the particular solution. Accordingly, I replace the constants c_1 and c_2 in the homogeneous solution with functions $A(t)$ and $B(t)$, to seek the particular solution of the form

$$x_p(t) = A(t) \cos \lambda t^2 + B(t) \sin \lambda t^2 \tag{5.13}$$

$$y_p(t) = -A(t) \sin \lambda t^2 + B(t) \cos \lambda t^2. \tag{5.14}$$

Substituting this into the equations of motion I now have:

$$\dot{A} \cos \lambda t^2 + \dot{B} \sin \lambda t^2 = 0, \tag{5.15}$$

$$-\dot{A} \sin \lambda t^2 + \dot{B} \cos \lambda t^2 = -\sigma; \tag{5.16}$$

therefore

$$\dot{A} = \sigma \sin \lambda t^2, \tag{5.17}$$

$$\dot{B} = -\sigma \cos \lambda t^2. \tag{5.18}$$

Eqs. (5.17) and (5.18) cannot be integrated analytically, but they can be expressed in terms of Fresnel integrals (Zwillinger, 1996). The Fresnel integrals are defined as follows:

$$S(t) = \int_0^t \sin t'^2 dt', \tag{5.19}$$

$$C(t) = \int_0^t \cos t'^2 dt'. \tag{5.20}$$

and have the following properties:

$$S(-t) = -S(t), \tag{5.21}$$

$$C(-t) = -C(t), \tag{5.22}$$

$$S(\infty) = C(\infty) = \sqrt{\frac{\pi}{8}}. \tag{5.23}$$

Therefore

$$A(t) = \frac{\sigma}{\sqrt{|\lambda|}} S\left(t\sqrt{|\lambda|}\right), \tag{5.24}$$

$$B(t) = -\frac{\sigma}{\sqrt{|\lambda|}} C\left(t\sqrt{|\lambda|}\right). \tag{5.25}$$

I denote initial conditions with a subscript i, and write the solution to Eqs. (5.9) and (5.10) as

$$x(t) = x_i \cos\left[\lambda\left(t^2 - t_i^2\right)\right] + y_i \sin\left[\lambda\left(t^2 - t_i^2\right)\right]$$
$$+ \frac{\sigma}{\sqrt{|\lambda|}}\left[(S - S_i)\cos\lambda t^2 - (C - C_i)\sin\lambda t^2\right], \tag{5.26}$$

$$y(t) = -x_i \sin\left[\lambda\left(t^2 - t_i^2\right)\right] + y_i \cos\left[\lambda\left(t^2 - t_i^2\right)\right]$$
$$- \frac{\sigma}{\sqrt{|\lambda|}}\left[(C - C_i)\cos\lambda t^2 + (S - S_i)\sin\lambda t^2\right]. \tag{5.27}$$

Because the asteroid is swept over by the secular resonance at time $t = 0$, I can calculate the changes in x, y by letting $t_i = -t_f$ and evaluating the coefficients C_i, C_f, S_i, S_f far from resonance passage, i.e., for $t_f\sqrt{|\lambda|} \gg 1$, by use of 5.23. Thus I find

$$x_f = x_i + \sigma\sqrt{\frac{\pi}{2|\lambda|}}\left[\cos\lambda t_i^2 - \sin\lambda t_i^2\right], \tag{5.28}$$

$$y_f = y_i - \sigma\sqrt{\frac{\pi}{2|\lambda|}}\left[\cos\lambda t_i^2 + \sin\lambda t_i^2\right]. \tag{5.29}$$

The new value of J after resonance passage is therefore given by

$$J_f = \frac{1}{2}\left(x_f^2 + y_f^2\right)$$
$$= \frac{1}{2}\left(x_i^2 + y_i^2\right) + \frac{\pi\sigma^2}{2|\lambda|} + \sigma\sqrt{\frac{\pi}{2|\lambda|}}[x_i(\cos\lambda t_i^2 - \sin\lambda t_i^2) - y_i(\cos\lambda t_i^2 + \sin\lambda t_i^2)]$$
$$= J_i + \frac{\pi\sigma^2}{2|\lambda|} + \sigma\sqrt{\frac{2\pi J_i}{|\lambda|}}\cos(\phi_i - \lambda t_i^2 - \frac{\pi}{4}). \tag{5.30}$$

With a judicious choice of the initial time, t_i, and without loss of generality, the cosine in the last term becomes $\cos\varpi_i$, and therefore

$$J_f = J_i + \frac{\pi\sigma^2}{2|\lambda|} + \sigma\sqrt{\frac{2\pi J_i}{|\lambda|}}\cos\varpi_i. \tag{5.31}$$

The asteroid's semimajor axis a is unchanged by the secular perturbation; therefore the changes in J due to the secular perturbation reflect changes in the asteroid's eccentricity e. Using the following definition:

$$\delta_e \equiv \left|\sigma\sqrt{\frac{\pi}{|\lambda|\sqrt{a}}}\right|, \tag{5.32}$$

For small e, I can use the approximation $J \simeq \frac{1}{2}\sqrt{a}e^2$. Considering all possible values of $\cos \varpi_i \in \{-1, +1\}$, the final and initial eccentricity are related by

$$e_{min,max} \simeq |e_i \pm \delta_e| . \tag{5.33}$$

For asteroids with non-zero initial eccentricity, the phase dependence in Eq. (5.33) means that secular resonance sweeping can potentially both excite and damp orbital eccentricities (depending on the particular value of δ_e). Considering all possible values of $\cos \varpi_i \in \{-1, +1\}$, the final eccentricity lies in the range $[|e_i - \delta_e|, |e_i + \delta_e|]$.

5.2 ν_6 sweeping of the Main Asteroid Belt

In order to apply Eqs. 5.31 and 5.33 to the problem of the ν_6 resonance sweeping through the asteroid belt, I must obtain values for the parameter σ (Eq. 5.3), for semimajor axis values in the main asteroid belt. I must also find the relationship between the giant planets orbits and the location of the ν_6 resonance. The location of the ν_6 resonance is defined as the semimajor axis, s_{ν_6}, where the rate, g_0 (Eq. 5.2), of pericenter precession of a massless particle (or asteroid) is equal to the g_6 eigenfrequency of the solar system. In the current solar system, the g_6 frequency is the secular mode with the most power in Saturn's eccentricity-pericenter variations, and is very sensitive to Saturn's semimajor axis. During the epoch of planetesimal-driven planet migration, Jupiter migrated by only a small amount but Saturn likely migrated significantly more (Fernandez and Ip, 1984; Malhotra, 1995; Tsiganis et al., 2005). I therefore adopt a simple model of planet migration in which Jupiter is fixed at 5.2 AU and only Saturn migrates. I also neglect any effects from the ice giants Uranus and Neptune, as well as secular effects due to the more massive icy planetesimal disk and more massive asteroid belt. In this simplified model, the g_6 frequency varies with time as Saturn migrates, so g_6 is solely a function of Saturn's semimajor axis. In contrast with the variation of g_6, there is negligible variation of the asteroid's precession rate, g_0, as Saturn migrates.

For fixed planetary semimajor axes, the Laplace-Lagrange secular theory provides a simple and direct way of calculating the secular frequencies of the planets.

In this theory, only the terms of the disturbing function are retained of second order in eccentricity and first order in mass (**Murray and Dermott, 1999**). In the planar two-planet case, the secular perturbations of planet j, where $j = 5$ is Jupiter and $j = 6$ is Saturn, are described by the following disturbing function:

$$R_j = \frac{n_j}{a_j^2}\left[\frac{1}{2}A_{jj}e_j^2 + A_{jk}e_5e_6\cos(\varpi_1 - \varpi_2)\right], \tag{5.34}$$

where n is the mean motion, and $\mathbf{A}$ is a matrix with elements

$$A_{jj} = +n_j\frac{1}{4}\frac{m_k}{M_\odot+m_j}\alpha_{jk}\bar{\alpha}_{jk}b_{3/2}^{(1)}(\alpha_{jk}), \tag{5.35}$$

$$A_{jk} = -n_j\frac{1}{4}\frac{m_k}{M_\odot+m_j}\alpha_{jk}\bar{\alpha}_{jk}b_{3/2}^{(2)}(\alpha_{jk}), \tag{5.36}$$

for $j = 5, 6$, $k = 6, 5$, and $j \neq k$; $\alpha_{jk} = \min\{a_j/a_k, a_j/a_k\}$, and

$$\bar{\alpha}_{jk} = \begin{cases} 1 & : a_j > a_k \\ a_j/a_k & : a_j < a_k. \end{cases} \tag{5.37}$$

The secular motion of the planets is then described by a set of linear differential equations for the eccentricity vectors, $e_j(\sin\varpi_j, \cos\varpi_j) \equiv (h_j, k_j)$,

$$\dot{h}_j = +\sum_{p=5}^{6} A_{pj}k_j, \quad \dot{k}_j = -\sum_{p=5}^{6} A_{pj}h_j. \tag{5.38}$$

For fixed planetary semimajor axes, the coefficients are constants, and the solution is given by a linear superposition of eigenmodes:

$$\{h_j, k_j\} = \sum_p E_j^{(p)}\{\cos(g_pt + \beta_p), \sin(g_pt + \beta_p)\}, \tag{5.39}$$

where g_p are the eigenfrequencies of the matrix $\mathbf{A}$ and $E_j^{(p)}$ are the corresponding eigenvectors; the amplitudes of the eigenvectors and the phases β_p are determined by initial conditions. In my 2-planet model, the secular frequencies g_5 and g_6 depend on the relative masses of Jupiter and Saturn to the Sun and on the semimajor axes of Jupiter and Saturn.

Using the current semimajor axes of Jupiter and Saturn the Laplace-Lagrance theory gives frequency values $g_5 = 3.7''$ yr^{-1} and $g_6 = 22.3''$ yr^{-1}, which are lower than the more accurate values given by **Brouwer and van Woerkom (1950)** by 14% and 20%, respectively (**Laskar, 1988**). **Brouwer and van Woerkom (1950)** achieved their more accurate solution by incorporating higher order terms in the disturbing function involving $2\lambda_5 - 5\lambda_6$, which arise due to Jupiter and Saturn's proximity to the 5:2 resonance (the so-called "Great Inequality"). However the effect of the 5:2 resonance is only important over a very narrow range in Saturn's semimajor axis. More significant is the perturbation owing to the 2:1 near-resonance of Jupiter and Saturn. **Malhotra et al. (1989)** developed corrections to the Laplace-Lagrange theory to account for the perturbations from $n + 1 : n$ resonances in the context of the Uranian satellite system. Applying that approach to my problem, I find that the 2:1 near-resonance between Jupiter and Saturn leads to zeroth order corrections to the elements of the **A** matrix. Including these corrections, I determined the secular frequencies for a range of values of Saturn's semimajor axis; the result for g_6 is shown in Fig. **5.1** (dashed line).

I have also calculated values for the eccentricity-pericenter eigenfrequencies by direct numerical integration of the two-planet, planar solar system. In these simulations, Jupiter's initial semimajor axis was 5.2 AU, Saturn's semimajor axis, a_{sat}, was one of 233 values in the range 7.3–10.45 AU, initial eccentricities of Jupiter and Saturn were 0.005, and initial inclinations were zero. The initial longitude of pericenter and mean anomalies of Jupiter were $\varpi_{jup,i} = 15°$ and $\lambda_{jup,i} = 92°$, and Saturn were $\varpi_{sat,i} = 338°$ and $\lambda_{sat,i} = 62.5°$. In each case, the planets orbits were integrated for 100 myr, and a Fourier transform of the time series of the $\{h_j, k_j\}$ yields their spectrum of frequencies. For regular (non-chaotic) orbits, the spectral frequencies are well defined and are readily identified with the frequencies of the secular solution. The result for the g_6 frequency obtained by this numerical analysis are shown by the solid line in Fig. **5.1** .

The comparison between the numerical analysis and the analytical solution indicates that the linear secular theory (with corrections for the 2:1 near-resonance) is an adequate model for the variation in g_6 as a function of a_{sat}. I therefore used the analytical secular theory to calculate the eigenvector components $E_j^{(6)}$ and the

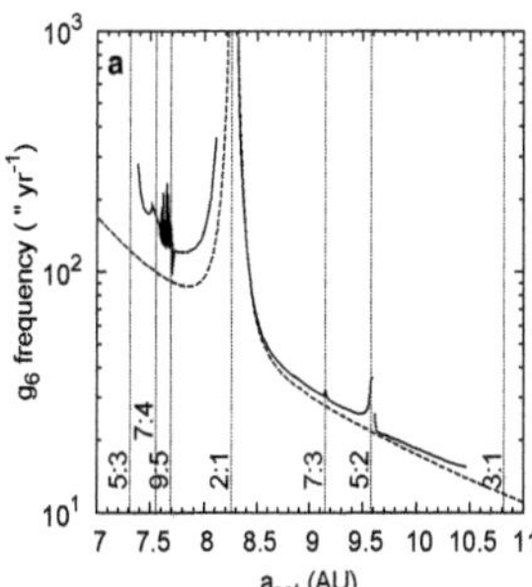
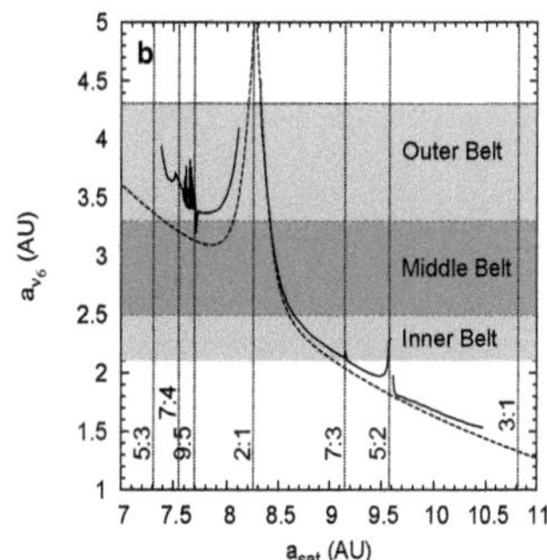

Figure 5.1: The ν_6 resonance during planet migration. **a)** The g_6 eigenfrequency as a function of Saturn's semimajor axes for Jupiter fixed at 5.2 AU. The dashed line shows the calculation using linear secular theory, with a correction for the effect of the near 2:1 mean motion resonance between Jupiter and Saturn (**Malhotra et al.**, 1989). The solid line was obtained from spectral analysis of 233 solar system integrations. The locations of Jupiter/Saturn MMRs which have an effect on the value of the g_6 in the range explored numerically are shown. The position of the ν_6 resonance (at zero inclination) as a function of Jupiter and Saturn's semimajor axes. **b)** The zero inclination position of the ν_6 resonance as a function of Saturn's position for Jupiter fixed at 5.2 AU. The dashed line shows the calculation using linear secular theory, with a correction for the effect of the near 2:1 mean motion resonance between Jupiter and Saturn (**Malhotra et al.**, 1989). The solid line was obtained by using the g_6 eigenfrequencies obtained from spectral analysis of the 233 n-body simulations from **a** and then calculating the positions where $g_0 - g_6 = 0$ (for a particle at zero inclination) at each a_{sat} location.

location a_{ν_6} of the ν_6. I used the same values for the initial conditions of Jupiter and Saturn as in the direct numerical integrations discussed above. The value of a_{ν_6} was found by solving for the value of a_0 such that $g_0 = g_6$, where g_0 was calculated using Eq. (**5.2**) and g_6 is the eigenfrequency associated with the $p = 6$ eigenmode. The results were used in Eq. (**5.3**) along with the value of a_{ν_6} as a function of a_{sat} from Fig. **5.1a**, and are plotted in Fig. **5.2**.

I checked the results of my analytical model against four different n-body simulations of test particles in the asteroid belt that experience the effects of a migrating Saturn. In each of the four simulations, 30 test particles were placed at 2.3 AU and

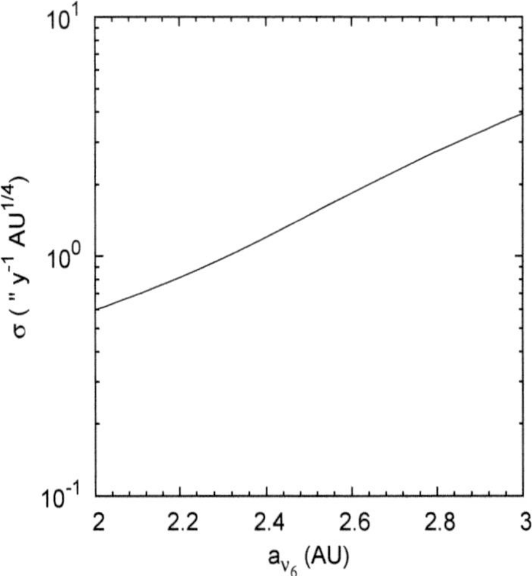

Figure 5.2: The value of the coefficient σ defined by Eq. (5.3) as a function of the zero inclination position of the ν_6 resonance. The value of $E_j^{(i)}$ as a function of Saturn's position (and hence on the position of the ν_6) was calculated using first order Laplace-Lagrange secular theory with corrections arising from the 2:1 Jupiter-Saturn mean motion resonance.

given different initial longitudes of pericenter spaced $10°$ apart. The numerical integration was performed using an implementation of a symplectic mapping (**Wisdom and Holman, 1991; Saha and Tremaine, 1992**). Jupiter and Saturn were the only planets included, and the asteroids were approximated as massless test particles. An artificial acceleration was applied to Saturn to cause it to migrate outward starting at 8.5 AU at the desired rate. The current Solar System values of the eccentricity of Jupiter and Saturn were used and inclinations were set to zero, and the time step was 0.01 yr. The only parameters varied between each of the four simulations were the initial osculating eccentricities of the test particles, e_i, and the migration speed of Saturn, $\dot{a}_{sat}$. The parameters explored were:

a) $e_i = 0.2$, $\dot{a}_{sat} = 1.0$ AU My^{-1};

b) $e_i = 0.2$, $\dot{a}_{sat} = 0.5$ AU My^{-1};

c) $e_i = 0.1$, $\dot{a}_{sat} = 1.0$ AU My^{-1};

d) $e_i = 0.3$, $\dot{a}_{sat} = 1.0$ AU My^{-1}.

Two forms of the analytical model were checked. First, the equations of motion given by Eqs. (5.9) and (5.10) were integrated using values for λ that were approximately equivalent to the values of $\dot{a}_{sat}$ used in the numerical integrations. Second, the eccentricity bounds calculated using Eq. (5.33) were calculated. The results of each of these four comparisons are shown in Fig. 5.3. While the test particles in the numerical integrations exhibit somewhat more complicated behavior than the analytical approximation, the values of the maximum and minimum final eccentricities from Eq. (5.33) (shown as horizontal dashed lines) are in good agreement with the final values of the eccentricities of the numerically integrated test particles. The slower sweep rates yield higher eccentricity excitation, but the phase dependence means that a secular resonance can both excite and damp eccentricities of asteroids that have a non-zero initial eccentricity. Eq. (5.33) somewhat underpredicts the maximum final eccentricity, which may be due to higher order terms in the disturbing function that may become more important at high eccentricity, as well as effects due to close encounters with Jupiter.

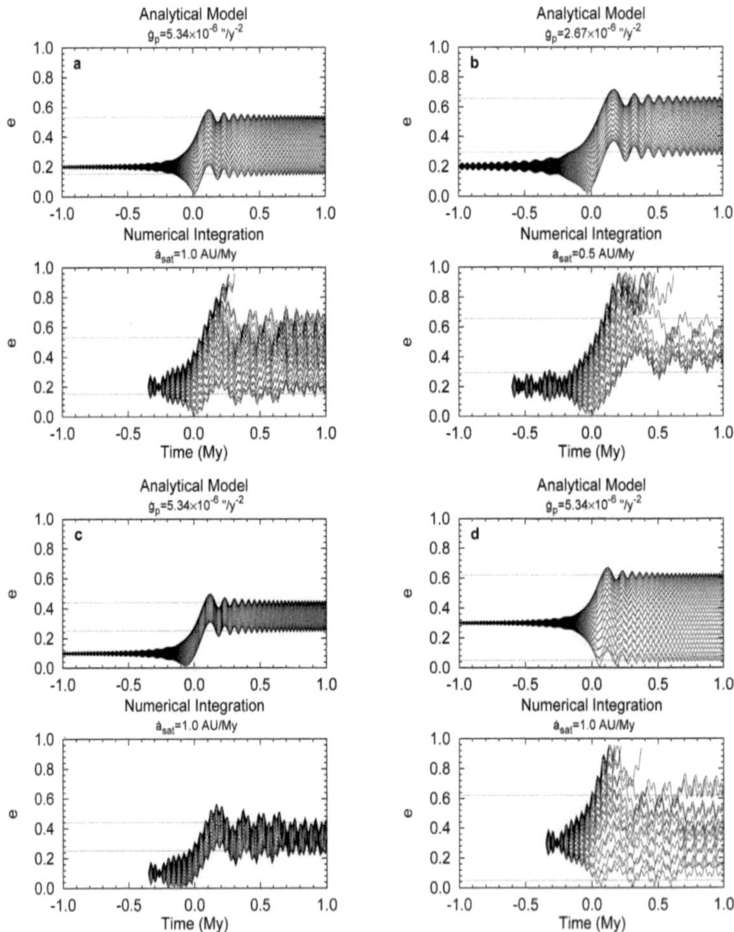

Figure 5.3: A comparison between the integrated equations of motion given by Eqs. (5.9) and (5.10) and n-body numerical integrations of test particles at 2.3 AU. The dashed lines represent the envelope of the predicted final eccentricity using Eq. (5.33).

5.3 Double-peaked asteroid eccentricity distribution

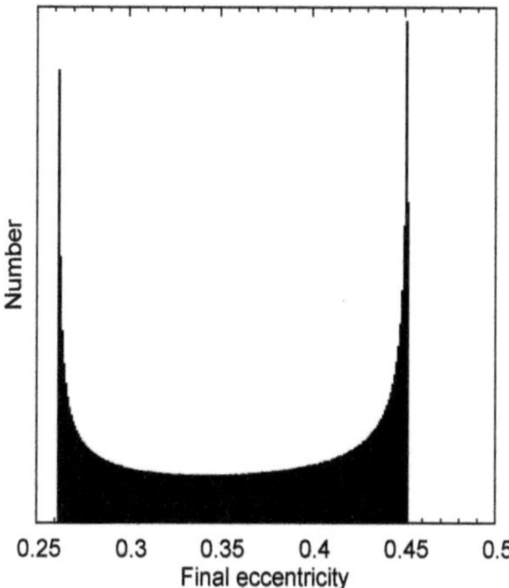

Figure 5.4: The final eccentricity distribution of an ensemble of particles with initial eccentricity $e_i = 0.1$ and uniformly distributed values of the phase angle ϖ_i. The effect due to the sweeping ν_6 resonance was modeled using Eq. (5.31). The parameters were chosen to simulate asteroids at $a = 2.3$ AU and $\dot{a}_{sat} = 1$ AU My^{-1}.

An important result of in Eq. (5.31) is that if the asteroid belt was initially "cold," that is asteroids were on nearly circular orbits prior to secular resonance sweeping, then the asteroids would be nearly uniformly excited to a narrow range of final eccentricities, the value of which would be determined by the rate of resonance sweeping. Secular resonance sweeping removes asteroids by exciting their eccentricities above planet-crossing values, therefore a uniformly excited asteroid belt will either lose all its asteroids or none. In order to have partial depletion of the asteroid belt as well as yield an eccentricity distribution with a significant dispersion, such as that seen in the observed asteroid belt shown in Fig. 5.5a, the asteroid eccentricities must be excited prior to resonance sweeping. However, the final eccentricities of an ensemble of test particles excited by the sweeping ν_6 resonance will not uniformly fill up the bounds given by Eq. (5.33). This is illustrated in Fig. 5.4 showing the effects of applying Eq. (5.31) to a simulated population of asteroids with an initial eccentricity $e_i = 0.1$ and a uniform distribution of the initial phase angle ϖ_i.

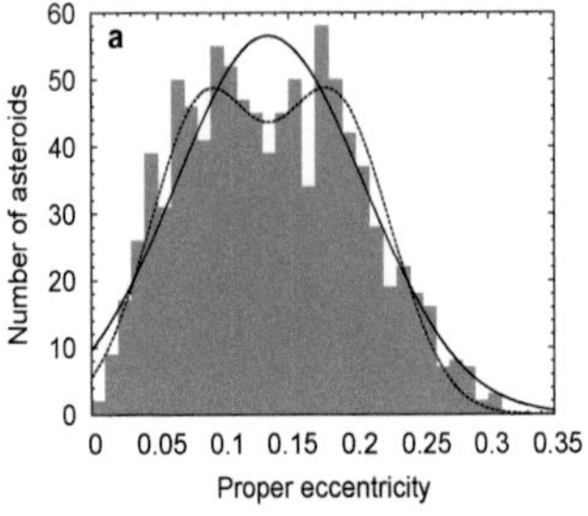
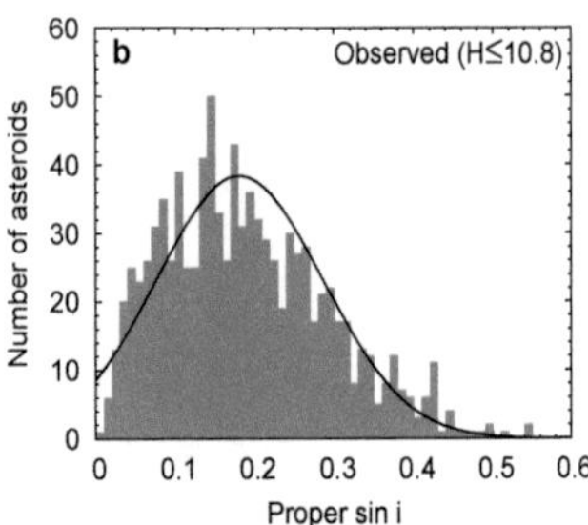

Figure 5.5: Proper eccentricity and inclination distributions of asteroids with $H \leq 10.8$. Only asteroids that were not identified as being members of collisional families are included here. The proper elements were taken from the AstDys online data service (Knežević and Milani, 2003). The solid lines are the best fit gaussian distributions to the observational data. The dashed line is the best fit double-peaked distribution for the eccentricity distribution.

When the simulated asteroids orbit with a range of semimajor axes from 2–2.8 AU, and have an initial eccentricity distribution rather than a single eccentricity, secular resonance sweeping produces a double-peaked eccentricity distribution. The results of applying Eq. (5.31) to an ensemble of simulated asteroids with a distribution of initial eccentricities and semimajor axes is shown in Fig. 5.6. Fig. 5.6a shows the initial eccentricity distribution, which is modeled as a Gaussian distribution with a mean at $e = 0.15$. Fig. 5.6b shows the same population after ν_6 resonance sweeping has occurred due to the migration of Saturn at a rate of 1.0 AU/ My. When an ensemble of asteroids with a Gaussian eccentricity distribution is subjected to the sweeping secular resonance, a double-peaked eccentricity distribution results. Because of the slight bias towards the upper limit of the eccentricity excitation band, proportionally more asteroids are found in the upper peak. If the high-e peak is above the Mars-crossing value, then that population will be eroded as planetary encounters remove the asteroids from the main belt. After erosion, the remaining asteroid belt may still exhibit a double peak, with the low-e peak corresponding to the original low-e peak produced by the ν_6 resonance sweeping, and the high-e

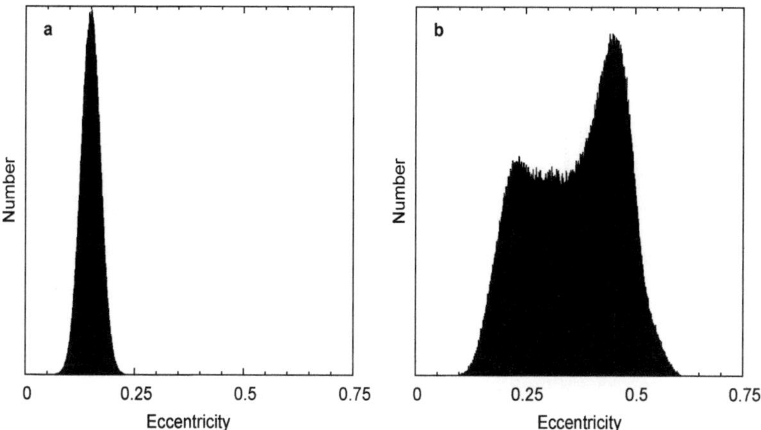

Figure 5.6: The effects of the sweeping ν_6 resonance on an ensemble of asteroids between 2–2.8 AU with a uniform distribution of phases. **a** The initial distribution of eccentricities. The mean of the distribution is 0.15 and the standard deviation is 0.02. **b** The final distribution of eccentricities after ν_6 sweeping, using the analytical model given by Eq. (5.31) with $\dot{a}_{sat} = 1$ AU My^{-1}. The y-axes are not to the same scale.

peak near the Mars-crossing eccentricity value. Alternatively, if the ν_6 resonance sweeping was very fast, the observed double peaked asteroid eccentricity distribution may be a direct consequence of the sweeping. In this case, very few asteroids would have been excited above the Mars-crossing value, and the asteroid belt may have experienced only moderate levels of depletion due to its new dynamical structure (i.e. the post-sweeping asteroid belt would contain new unstable zones at the present-day locations of mean motion and secular resonances). In either case, the observed double-peaked asteroid eccentricity distribution is well explained by the sweeping of the ν_6 secular resonance.

5.4 A Constraint on Saturn's migration rate

By relating the g_6 secular frequency to the semimajor axis of Saturn, $\dot{g}_6$ can be related to the migration rate of Saturn, $\dot{a}_{sat}$. I used the results of my analytical model to set limits on the rate of migration of Saturn, with the caveat that many important effects are ignored, such as asteroid-Jupiter mean motion resonances, and Jupiter-Saturn mean motion resonances with the exception of the 2:1 resonance. I have confined my analysis to only the inner main belt, between 2.2–2.8 AU. Beyond 2.8 AU strong jovian mean motion resonance become more numerous. The migration of Jupiter would have caused strong jovian mean motion resonances to sweep the asteroid belt, causing additional depletion beyond that of the sweeping ν_6 resonance. There is evidence from the distribution of asteroids in the main belt that the sweeping of the 5:2, 7:3, and 2:1 jovian mean motion resonances may have depleted the main belt (**Minton and Malhotra, 2009**). A further complication is that sweeping jovian mean motion resonances may have also trapped icy planetesimals that entered the asteroid belt region from their source region beyond Neptune (**Levison et al., 2009**). The effects of these complications are reduced when I only consider the inner asteroid belt. From Fig. **5.1b**, the ν_6 would have swept the inner asteroid belt region between 2.2–2.8 AU when Saturn was between ~ 8.5–9.2 AU. Therefore the limits on $\dot{a}_{sat}$ that I set using the inner asteroid belt as a constraint are only applicable for this particular portion of Saturn's migration history.

An estimated final eccentricity as a function of initial asteroid semimajor axis, initial asteroid eccentricity, and the migration rate of Saturn is shown in Fig. **5.7**. The larger the initial asteroid eccentricities, the wider the bounds in their final eccentricities. If I adopt the criterion that an asteroid is lost from the main belt when it achieves a planet-crossing orbit (either Jupiter or Mars) and that initial asteroid eccentricities were therefore confined to $\lesssim 0.4$, then from Fig. **5.7** Saturn's migration rate while the ν_6 resonance was passing through the inner asteroid belt must have been $\dot{a}_{sat} \gtrsim 0.2$ AU My^{-1}. This model suggests that if Saturn's migration rate had been slower than 0.2 AU My^{-1} while it was migrating across ~ 8.5–9.2 AU, then the inner asteroid belt would have been completely swept clear of asteroids by the ν_6 resonance.

The observed eccentricity distribution of the main asteroid belt may be used to further constrain the migration rate of Saturn. As Fig. **5.5** shows, the eccentricity

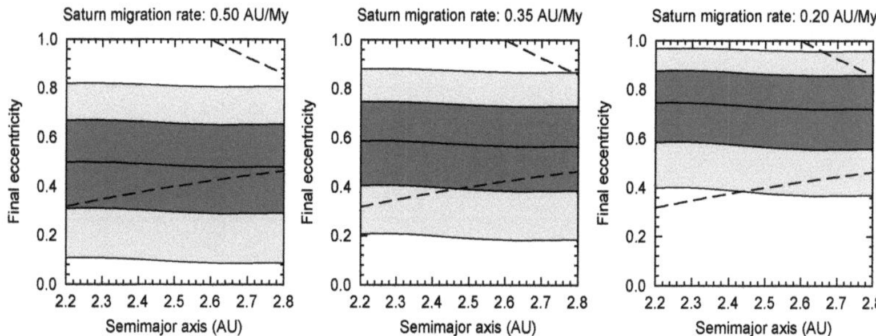

Figure 5.7: Estimated final eccentricity of asteroids as a function of asteroid semimajor axis and eccentricity for three different migration rates of Saturn using Eq. (5.33). Asteroids swept by the ν_6 resonance can have a range of final eccentricities depending on their apsidal phase, ϖ_i. The outermost shaded region demarcates the range of final eccentricities for asteroids with an initial eccentricity $e_i = 0.4$. The innermost shaded region demarcates the range of final eccentricities for asteroids with an initial eccentricity $e_i = 0.2$. The solid line at the center of the shaded regions is the final eccentricity for an asteroid with an initial eccnetricity $e_i = 0$.

distribution of large, primordial asteroids is well modeled as a double-peak Gaussian. If the pre-sweeping asteroid belt had a Gaussian eccentricity distribution, then the lower peak of the post-sweeping asteroid belt should be equal to the lower bound of Eq. (5.33) because the peaks of the post-sweeping distribution correspond to the upper and lower bounds of Eq. (5.33) for e_i equal to the peak eccentricity of the pre-sweeping distribution (note that the equation relates single e_i values to e_f values, but here I am applying it to eccentricity distributions). Using Eq. (5.33), assuming that the pre-sweeping eccentricity distribution had a peak at 0.4 and very little dispersion (the maximum possible value), then the low-e peak of the eccentricity distribution at ~ 0.09 implies that $\dot{a}_{sat} \gtrsim 0.5$ AU My^{-1}. If the peak of the initial eccentricity distribution was smaller than 0.4 then the lower limit of $\dot{a}_{sat}$ would be even higher. The lower limit on the rate of Saturn's migration as a function of Saturn's semimajor axis is plotted in Fig. 5.8 (the upper horizontal line). The lower limit of $\dot{a}_{sat} > 0.2$ AU My^{-1} that would result in all primordial asteroids being excited to Mars-crossing orbits, regardless of initial eccentricity, is also plotted in

Fig. 5.8 (the lower horizontal line).

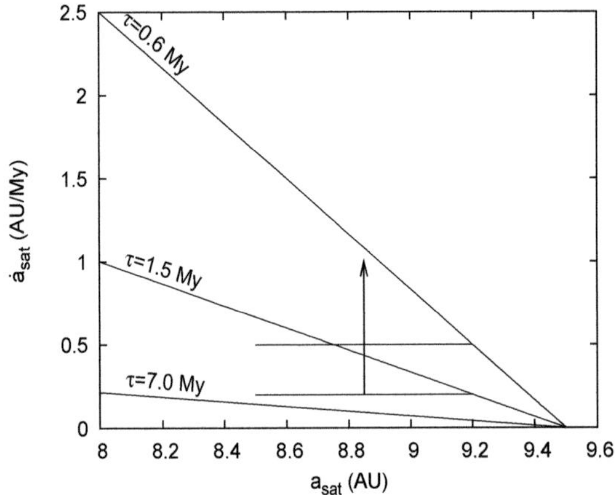

Figure 5.8: Limits on the migration rate of Saturn as a function of Saturn's semi-major axis. The horizontal lines are the lower limits of $\dot{a}_{sat}$ derived here. The lower horizontal line corresponds to the limit $\dot{a}_{sat} > 0.2$ AU My^{-1}. If Saturn had migrated at a rate slower than this limit, then all asteroids swept by the ν_6 resonance would have been excited to Mars-crossing orbits. The upper horizontal line corresponds to the limit $\dot{a}_{sat} > 0.5$ AU My^{-1}. At this limit, an asteroid belt with an initially Gaussian eccentricity distribution with a mean of ~ 0.4 would have a double peak distribution after being swept by the ν_6 resonance with the lower peak corresponding to the lower peak of the best fit double Gaussian distribution to the observed asteroids shown in Fig. 5.5a. The solid diagonal lines are the migration rate as a function of semimajor axis using the migration model given by Eq. (5.40) for three different values of the timescale τ.

Some studies of planet migration model the migration history of the giant planets as exponential functions, of the form:

$$a = a_0 + \Delta a(1 - e^{-t/\tau}),\qquad(5.40)$$

where a_0 is the initial semimajor axis of the planet, Δa is the migration distance, and τ is a time scale. For comparison I have plotted model migration paths for three different values of the timescale τ in Fig. 5.8. The timescale $\tau = 0.6$ My is that needed to always be above the rate limit set by the distribution of asteroids in the main belt, the most stringent rate limit discussed here. The timescale $\tau = 1.5$ My is that needed to always be above the lower rate limit of $\dot{a}_{sat} > 0.5$ AU My^{-1}.

5.5 Summary and Discussion

My results suggest that the initial dynamical excitation of the asteroid belt during the epoch of planetesimal-driven giant planet migration was relatively short-lived. Based on the analytical model of the sweeping ν_6 resonance developed here, if Saturn's migration rate must have been $\dot{a}_{sat} \gtrsim 0.1$ AU My^{-1} while Saturn migrated between 8.5–9.2 AU. If Saturn had migrated slower than this, then the inner asteroid belt would look quite different today than the one that we observe. My analytical model predicts that the action of the ν_6 resonance should have excited all asteroids into planet-crossing orbits if Saturn had migrated at such a slow speed. This lower limit on the migration rate of Saturn can be further refined by noting a unique feature of the asteroid belt's eccentricity distribution. Based on the observed double-peaked eccentricity distribution of the population of asteroids with $H \leq 10.8$, I estimate that Saturn's migration rate must have been $\dot{a}_{sat} \gtrsim 0.5$ AU My^{-1} while Saturn migrated between 8.5–9.2 AU. These lower limits on the migration rate implies that Saturn's total migration may have taken place in $\sim$ 1–10 My, or if an exponential model of planet migration is used such as that of Eq. (5.40), the migration timescale must have been $\tau \lesssim 1.5$–0.6 My. However, in my analysis I have neglected the effects of sweeping mean motion resonances due to the migration of Jupiter. Mean motion resonance sweeping may reduced the efficiency of secular resonance sweeping. Asteroids temporarily trapped inside a sweeping mean motion resonance would have pericenter precession rates that are different than those calculated using Eq. (5.2). However the observed semimajor axis distribution of the asteroid belt would be expected to be very different than that of the observed belt if the only asteroids that survived the passage of the ν_6 resonance were those that were temporarily trapped in mean motion resonances. Under this scenario, the Jupiter-facing sides of mean motion resonances would be expected to have an excess of asteroids compared with the Sun-facing sides, because only the Jupiter-facing sides of the modern-day Kirkwood gaps would have felt the presence of sweeping strong mean motion resonances during the sunward migration of Jupiter. This would be in contradiction to the observed asteroid belt where the observed the Jupiter-facing sides of mean motion resonances show a depletion of asteroids (**Minton and Malhotra, 2009**).

These results depend strongly on the eccentricity of the giant planets during the migration. The coefficient σ is proportional to the power of the g_6 mode, which is related to the eccentricities of the giant planets (namely Saturn and Jupiter). Therefore if the giant planets' orbits were more circular than they are today while the ν_6 resonance was passing through the inner asteroid belt, then the migration timescale could have been longer without having cleared away all asteroids. From Eq. (**5.39**), and the definition $e_j(\sin \varpi_j, \cos \varpi_j) \equiv (h_j, k_j)$, the value of $E_j^{(p)}$ is a linear combination of the eccentricities of the giant planets. Because Saturn is the planet with the largest amplitude of the $p = 6$ mode, from Eq. (**5.31**) the relationship between the sweep rate and the value of Saturn's eccentricity is approximately $\lambda_{min} \propto e_{sat}^2$. Therefore, to increase the limiting timescale by a factor of ten would only require that the giant planets' eccentricities were $\sim 0.3\times$ their current value (or $e_{jup,sat} \approx 0.015$). However, this would need to have occurred while Saturn was between ~ 8.5–9.2 AU, so some mechanism would need to have increased Saturn's eccentricity up to its present value after the ν_6 had already swept through the inner asteroid belt. One solution to this dilemma may be that Eq. (**5.40**) is simply a poor model for the migration history of the giant planets. A model in which Saturn's migration rate remained relatively constant throughout the era planet migration rather would last longer than an exponential model. For example, a migration rate limit of $\dot{a}_{sat} > 0.5$ AU My^{-1} means that the total time of migration is $t_{mig} \lesssim 2$ My assuming that Saturn migrated outward a distance of ~ 1 AU.

The short timescale derived here is in disagreement with other published estimates of the migration rate of the giant planets. **Murray-Clay and Chiang (2005)** exclude $\tau \leq 1$ My to 99.65% confidence using the lack of a large observed asymmetry in the population of Kuiper belt objects in the two libration centers of the 2:1 Neptune mean motion resonance. **Boué et al.** (2009) exclude $\tau \leq 7$ My based on the observed tilt of Saturn. They find that $\tau > 7$ My, and the migration path using this timescale limit is also plotted on Fig. 5.8. My results suggest that if planet migration occurred at the slow rates implied by these observations, then the inner asteroid belt should have been swept clear of asteroids by the effect of the ν_6 secular resonance. **Brasser et al.** (2009) have suggested that no smooth migration model can account for the currently observed secular amplitudes in the giant planet system, and that a giant planet scattering event must have taken place. Giant planet scattering could potentially cause the giant planets to migrate in timescales comparable to those found here. Resolving these disparate giant planet migration timescales will be a goal for future models of planet migration. In addition, better modeling of the combined effects of overlapping sweeping mean motion and secular resonances will better explain how some asteroids can survive when Saturn's migration rate is slower than the lower limits derived here. Understanding the nature of the observed double-peaked eccentricity distribution may also help illuminate the pre-migration state of the asteroid belt, which may help constrain mechanisms for producing the primordial excitation and level of depletion of the main belt. Applying the analytical model developed here for the ν_6 resonance to the ν_{16} resonance (an inclination-longitude of the ascending node secular resonance) will also be able to take advantage of the observed inclination distribution of the main belt as a further constraint on the migration history of the giant planets.

CHAPTER 6

SOLAR WIND LITHIUM ENHANCEMENT BY PLANETESIMAL BOMBARDMENT

6.1 Introduction

The solar photosphere is highly depleted in lithium relative to the chondritic abundance, with the photosphere depleted in Li by a factor of 170 ± 60 relative to chondrites (**Asplund et al., 2006**). The two stable isotopes of lithium are destroyed in the Sun at different rates, with ^{6}Li being destroyed $\sim$90$\times$ faster than ^{7}Li (**Proffitt and Michaud, 1989**). Since lithium is the only element that is significantly enriched in chondrites relative to the photosphere, the in-fall of material with a chondritic ^{6}Li/^{7}Li ratio, such as asteroids or comets, into the Sun may have had an effect on the ^{6}Li/^{7}Li ratio in the solar photosphere. The composition of the solar wind is expected to reflect the composition of the solar photosphere, therefore a measurement of the lithium isotope ratio in solar wind implanted material could be useful in understanding the bombardment history of the inner solar system.

The current asteroid belt contains $\sim 6 \times 10^{-4}$ $M_\oplus$ of material (**Krasinsky et al., 2002**). Evidence from lunar and meteoritic samples, as well as the cratering record of the inner terrestrial planets, seems to indicate that the asteroid belt suffered a dynamical depletion event $\sim$ 500 My after the planets formed. This event has been termed the Lunar Cataclysm, or more generally the Late Heavy Bombardment (LHB), and it resulted in a surge in the impact rate in the inner solar system and lasted between 50-200 My (**Strom et al., 2005**).

The amount of material that was removed from the asteroid belt during the LHB is poorly constrained, and the current best estimates place the pre-LHB main belt mass at about ~ 10–$20\times$ its current mass, or roughly 6–12×10^{-3} $M_\oplus$ (**Levison et al., 2001; Bottke et al., 2005b; O'Brien et al., 2007**). In addition to asteroids, comets may also have been responsible for some of inner solar system impactors. It is also not known how much cometary material was involved in the LHB, but the size distribution of impactors matches the size distribution of the main asteroid belt, implying that at least the end of the LHB was dominated by asteroids and that there was no size selection during the depletion of the asteroid belt (Strom et al., 2005). Models of planetesimal-driven planet migration suggest that the influx of comets onto the terrestrial planets during the migration-induced perturbation would have roughly equalled that of asteroids (Gomes et al., 2005).

The asteroids delivered to the inner solar system by the LHB were likely excited by the sweeping of resonances as Jupiter and Saturn migrated. The two most important resonances for delivering main belt asteroids to the inner solar system are the ν_6 secular resonance and the 3:1 mean motion resonance with Jupiter. About ~ 65–75% of objects excited by these resonances impact the Sun (Gladman et al., 1997; Ito and Malhotra, 2006). However, as I showed in Chapter 2, §2.5.1, the overall probability of solar impact from the main asteroid belt is $\sim 15\%$.

In this paper I will attempt to develop an analytical model of the solar wind $^6\mathrm{Li}/^7\mathrm{Li}$ ratio both before, during, and after the LHB. In order to properly model this quantity the following questions need to be answered:

a) How much lithium did the Sun have at the time of the LHB?

b) What were the nuclear reaction rates of each lithium isotope in the solar convective zone at the time of the LHB?

c) What was the mass flux of the solar wind at the time of the LHB?

d) Are there alternative sources of solar wind $^6\mathrm{Li}$?

e) How much chondritic material impacted the Sun during the LHB?

f) How much dust was produced during the LHB, and what fraction of it was ionized and directly injected into the solar wind rather than accreted onto the solar convective zone?

If reasonable quantitative answers to each of the above questions can be found then it may be possible to answer the question: Would we be able to see evidence for the LHB in the form of an isotopic anomaly in solar wind-implanted lithium in lunar regolith that was exposed at 4 Ga?

6.2 The history of lithium in the solar system

Lithium depletion is an important but contentious issue in stellar evolution studies. Lithium is the only "metal" produced in the Big Bang in any significant amount, and it is destroyed by thermonuclear reactions in stars at relatively low temperatures, with ^{7}Li burning above $\sim 2.6 \times 10^6$ K, and ^{6}Li burning above $\sim 2.2 \times 10^6$ K (Jeffries, 2000; Proffitt and Michaud, 1989). However, the solar convective zone does not reach temperatures high enough to initiate any lithium burning. Therefore standard solar models have traditionally predicted that the majority of the lithium depletion occurred during the pre-main-sequence stage in order to account for the currently measured level of depletion (Piau and Turck-Chièze, 2002; Sestito et al., 2006; Umezu and Saio, 2000).

Studies of G types stars in young stellar clusters seems to indicate that little, if any, pre-main-sequence lithium depletion occurs for a ~ 1 M$_\odot$ star, which is in disagreement with the standard solar models. Therefore there must be some other mechanism responsible for depleting the Sun in lithium (Randich et al., 2001; Jeffries, 2000; Pasquini, 2000; Piau et al., 2003). The currently favored hypothesis is that there is mixing at the boundary between the solar convective and radiative zones induced by the Sun's rotation, and lithium in the convective zone is mixed down to temperatures where it can be destroyed (Deliyannis, 2000; Umezu and Saio, 2000). Therefore, for the analytical model I will develop in this work I'll assume that the Sun's zero age main sequence (ZAMS) lithium abundance is the chondritic value, and all depletion occurs during main sequence evolution.

6.2.1 Lithium destruction rates in Sun-like stars

Studies of sun-like stars in the Hyades cluster, with an age of $\sim$600 My, indicate that a lithium depletion of about $^7\text{Li}_0/^7\text{Li} \approx 3 - 5$ occurs in the first 500 My from ZAMS, with the rest occurring afterwards (**Piau et al., 2003**). This suggests, based on current solar and chondritic abundances of lithium, that the Sun had an average lithium destruction e-folding time of $\tau_d = 380 \pm 70$ My prior to 500 My in age and a subsequent average e-folding time of $\tau_d = 1.1 \pm 0.2$ Gy afterwards. In my analytical model, a range of different e-folding times is considered that reproduce the current Sun's lithium abundance.

Since ^6Li destruction occurs at a somewhat lower temperature than ^7Li destruction, its depletion rate is much greater. **Proffitt and Michaud (1989)** proposed a simple analytical relationship between the abundances of the two isotopes of lithium and their relative destruction rates:

$$\left(\frac{^6\text{Li}}{^6\text{Li}_0}\right) = \left(\frac{^7\text{Li}}{^7\text{Li}_0}\right)^r , \tag{6.1}$$

where r is the ratio of the ^6Li destruction time to the ^7Li destruction rate. The nuclear reaction rate is highly temperature dependant, but according to **Proffitt and Michaud**, $r \approx 90$ is a good estimate for the Sun.

6.2.2 The solar wind composition and time-varying mass flow rate.

It is thought that the solar wind composition is nearly identical to the composition of the solar convective zone. Therefore the solar wind mass flux of some species n can be given by:

$$\dot{M}_{n,s} = \dot{M}_s \cdot M_{fcz,n}, \tag{6.2}$$

where $M_{fcz,n}$ is the mass fraction of species n in the solar convective zone, and $\dot{M}_s$ is the solar wind mass flux.

The solar wind mass flux is currently $\dot{M}_s \approx 2 - 3 \times 10^{-14}$ M$_\odot$ yr^{-1} **Wood et al. (2005); Sackmann and Boothroyd (2003)**. Since the Sun is thought to be a typical G type dwarf star, a mass loss rate history of the Sun could be inferred from measuring stellar outflows of other sun-like stars during various stages of main sequence evolution. Wood et al. (2005) have studied stellar wind fluxes of several solar-type main sequence stars. By measuring the amount of Lyα absorption in a star, and fitting the observed absorption feature to a model astrosphere interacting with its local interstellar medium, the mass loss rate of the star due to stellar winds can be estimated (Zank, 1999). In order to use this method to determine the rate of mass loss of a measured star due to its stellar wind, the density and relative velocity of the local interstellar medium with respect to the star must also be known. Using the empirical formula of Wood et al. the solar wind mass flux as a function of age, t, can be estimated to be:

$$\dot{M}_s = \dot{M}_\odot \left(\frac{t}{t_\odot}\right)^{-2.33\pm0.55}, \quad t \gtrsim 400 - 700 \text{ My}, \tag{6.3}$$

where $\dot{M}_\odot = 1.6 \pm 0.9 \times 10^{12}$ g s^{-1} is the current mass loss rate of the Sun due to solar wind, and $t_\odot = 4.56$ Gy is the current age of the Sun (Dearborn, 1991). Prior to $400 - 700$ My the solar wind mass loss is comparable to the current solar wind. This implies that if the Sun had reached the transition from low to high solar wind mass loss then the solar wind mass flux could have been 50-600$\times$ higher at the time of LHB than it is currently. However since the transition time is concurrent with the LHB, both the low and high solar wind mass fluxes will be considered here in this model.

6.2.3   The solar wind-implanted lunar soil ^{6}Li anomaly

Since lithium, unlike noble gases, is naturally abundant in rocks, measuring a solar wind component of lithium is difficult. However, recently **Chaussidon and Robert** (1999) have measured solar wind-implanted lithium in lunar soils, giving an isotope ratio of ^{6}Li/^{7}Li$= 3.2 \pm 0.4 \times 10^{-2}$, compared to the chondritic ratio of ^{6}Li/^{7}Li$= 8.24 \pm 0.2 \times 10^{-2}$. This is in disagreement with standard solar convection models that predict efficient destruction of the lighter isotope of lithium, which should leave a photospheric ratio of ^{6}Li/^{7}Li $< 10^{-6}$ (**Chaussidon and Robert**, 1999). The proposed mechanism for the enhancement of ^{6}Li in the solar wind is nucleosynthesis in solar flares (**Ramaty et al.**, 2000). In order to account for the current ^{6}Li/^{7}Li ratio measured from lunar soils by **Chaussidon and Robert**, the rate of ^{6}Li production in the solar corona must be 2.4 ± 0.7 g s^{-1}. Since the lithium isotope measurement was made in lunar soils, which are expected to be overturned on relatively short timescales due to impact gardening, the measured ^{6}Li/^{7}Li ratio can probably be assumed to be the current value in the solar wind.

6.3 Estimating the rate of impacts on the Sun

Currently the upper layers of the Sun receive interplanetary material in a number of ways. The most obvious is due to impacts of asteroids and comets on highly eccentric orbits. Also, interplanetary dust particles (IDPs) larger than $\sim$0.5 μm spiral into the Sun due to the Poynting-Robertson effect. Comets and asteroids with perihelia very near to the Sun can become "sungrazers," shedding material that can rapidly become ionized and carried away as part of the solar wind (**Bzowski and Królikowska**, 2005). Dust in-fall is likely more important than the flux of large objects impacting the Sun in enhancing chondritic abundances of lithium in the solar wind because of the lower density of the solar wind compared with the solar convective envelope.

A simple estimate of the enhancement of solar wind ^{6}Li/^{7}Li can be made taking into account the sources of chondritic material as well as the mixing and destruction of lithium in the solar convective envelope. In this model, it is assumed that the mass flux of chondritic material that impacts the Sun directly, $\dot{M}_{in}$, is constant before, during, and after the LHB. I then divide the solar wind abundances into two sources corresponding to a source derived from the solar convective zone ($\dot{M}_s$) and a source corresponding to the injection of sublimated and ionized dust directly into the solar wind ($\dot{M}_d$).

Assuming $\sim 15\%$ of asteroids destabilized during the LHB impact the Sun, this gives a total deposited mass of $1.3 \pm 0.5 \times 10^{-3}$ M$_\oplus$. I will assume that the cometary flux is equalt to that of the asteroid flux. The cometary flux will be modeled as a step-function impact rate profile, that is the comet impact mass flux, $\dot{M}_{com,in}$, is simply the total deposited mass divided by the length of time of the epoch of planetesimal-driven planet migration. From Chapter **5** I showed that this time period could be no longer thatn 10 My. The asteroidal impact flux will be modeled using the results of Sim 3 of Chapter **2**, as shown in Fig. **2.11**.

An estimate of the asteroidal mass flux into the Sun before and after the LHB can be made by considering the mass loss rate of objects from the Main Asteroid Belt into the Near Earth Asteroid (NEA) population, assuming $\sim 15\%$ of the NEA objects ultimately impact the sun. This mass loss rate has been estimated to be $\sim 10^{-8}$ $M_\oplus$ My^{-1} (**Bottke et al., 2002b**). Since the mass of the asteroid belt was presumably 10 times higher prior to the LHB, the mass loss rate would have been correspondingly higher at that time.

The impactors in the analytical model developed here have chondritic abundances and isotopic ratios of lithium. There is some uncertainty regarding solar abundances of elements. The chondritic lithium abundance is approximately 170 ± 60 solar (**Grevesse and Sauval, 1998; Asplund et al., 2006**). The chondritic lithium isotopic ratio is $^6Li/^7Li \approx 0.0824$ (**Chaussidon and Robert, 1998**). In the model, impactors simply add their mass of lithium to the convective zone. The implanted solar wind lithium isotopic ratio can be given by the following:

$$\left(\frac{^6Li}{^7Li}\right)_{imp} = \frac{\bar{m}_{7Li}}{\bar{m}_{6Li}} \left(\frac{\dot{M}_{6Li,s} + \dot{M}_{6Li,d}}{\dot{M}_{7Li,s} + \dot{M}_{7Li,d}}\right) \tag{6.4}$$

where $\dot{M}_n$ is the mass flux into the solar wind of species n. The dust contribution of elements is:

$$\dot{M}_{n,d} = \dot{M}_d \cdot M_{f,Si} \cdot \frac{[n]}{[Si]} \cdot \frac{\bar{m}_n}{\bar{m}_{Si}}, \tag{6.5}$$

where $M_{f,Si}$ is the mass fraction of silicon in the dust (~ 0.15), $[n]/[Si]$ is the measured chondritic abundances of species n relative to silicon, and $\bar{m}_n/\bar{m}_{Si}$ is the ratio of atomic weights of species n and silicon.

For objects that impact the Sun, the material is mixed into the solar convective zone in a time comparable to the convective overturn time τ_c. The convective overturn time for the Sun can be estimated using the empirical formula given by **Noyes** et al. (1984):

$$\log(\tau_c) = 1.362 - 0.14x, \ x < 0, \tag{6.6}$$

where $x = 1 - (B - V)$ and τ_c has units of days. The convective overturn time is on the order of a few weeks, which is substantially shorter than the timescales considered for the LHB. Therefore I will assume that objects that impact the Sun are instantaneously mixed into the convective zone.

The total mass of lithium in the convection zone at any time t beyond some reference time $t_0 = 0$, results from a balance between lithium addition from meteoritic in-fall at a rate $\dot{M}_{in}$ and its destruction with a characteristic time τ_d. This balance can be expressed by the following differential equations:

$$\dot{M}_{^7\mathrm{Li,cz}} = -\frac{M_{^7\mathrm{Li,cz}}}{\tau_d} + \dot{M}_{^7\mathrm{Li,in}} \tag{6.7}$$

$$\dot{M}_{^6\mathrm{Li,cz}} = -\frac{M_{^6\mathrm{Li,cz}}}{\tau_d/r} + \dot{M}_{^6\mathrm{Li,in}} \tag{6.8}$$

Solving these two equations assuming constant meteoritic in-fall rates yields:

$$M_{^7\mathrm{Li,cz}} = \left(M_{^7\mathrm{Li,cz}}(0) - \dot{M}_{^7\mathrm{Li,in}}\tau_d \right) e^{-t/\tau_d} + \dot{M}_{^7\mathrm{Li,in}}\tau_d \tag{6.9}$$

$$M_{^6\mathrm{Li,cz}} = \left(M_{^6\mathrm{Li,cz}}(0) - \dot{M}_{^6\mathrm{Li,in}}\tau_d/r \right) e^{-rt/\tau_d} + \dot{M}_{^6\mathrm{Li,in}}\tau_d/r \tag{6.10}$$

The mass fraction of the two lithium isotopes in the Sun is:

$$M_{fcz,^7\mathrm{Li}} = \frac{M_{^7\mathrm{Li,cz}}}{M_{cz}} \tag{6.11}$$

$$M_{fcz,^6\mathrm{Li}} = \frac{M_{^6\mathrm{Li,cz}}}{M_{cz}} \tag{6.12}$$

The depth of the solar convective zone has been measured using helioseismology to be $R_{CZ} = 0.713 \pm 0.001 \ \mathrm{R}_\odot$ (**Basu and Antia, 1997**). According to standard solar models, the convective zone of the Sun contains a mass fraction of the total Sun of $M_{cz}/\ \mathrm{M}_\odot = 0.0241$ (**Bahcall et al., 2005, 2006**). This corresponds to a total mass of the convective zone of $M_{cz} = 4.8 \times 10^{31}$ g.

6.4 Dust production rates and the ultimate fate of dust in the solar system

Dust production may have a very important effect on the composition of solar wind particles. Silicate dust particles that reach inward of $\sim$20 R$_\odot$ from the Sun begin to sublimate (**Mann et al., 2004**). The sublimated atoms can then become ionized due to charge exchanges with solar wind protons, and subsequently carried away with the solar wind as pickup ions (PIUs) (**Bzowski and Królikowska, 2005**). Since at any given time the solar wind contains much less mass than the solar convective zone, chondritic material that is directly injected into the solar wind will have a much stronger effect on the solar wind composition than the same amount of material impacting the Sun and being diluted in the convection zone.

Another important source of near-solar dust are comets and asteroids with extremely low perihelia, which are called "sungrazer" comets (**Biesecker et al., 2002**). Currently the PIU production rates of sungrazer comets has been estimated to be $\sim 3 \times 10^4$ g s^{-1}, with the majority coming from the Kreutz family of comets (**Bzowski and Królikowska, 2005**). This corresponds to $\sim$1% of the current estimated in-fall rate of asteroids into the Sun using NEA population replenishment rates given by Bottke et al. (2002b). Due to the nature of the ν_6 secular resonance and 3:1 Jovian mean motion resonance resonance, most asteroids destabilized during the LHB would attain very high eccentricities before impacting the Sun (**Ito and Malhotra, 2006**). This could possibly have created a large number of sungrazing asteroids that produced prodigious amounts of near-solar dust. Collision rates in the asteroid belt are currently low, but during the LHB dynamical event collision rates could have been higher, also increasing the amount of interplanetary dust.

Sekanina (2003) estimated the total mass of currently active sungrazing comet families to be $\sim 10^{19}$ g. I'll make the simple assumption that the rate of PIU production is directly related to the total size of active sungrazers. Therefore:

$$\frac{\dot{M}_{PIU}}{M_{obj}} \approx 3 \times 10^{-15} \text{ s}^{-1} \tag{6.13}$$

This can be used to estimate the PIU production rate of any active sungrazing comet or asteroid. This implies that the average PIU production over time is dependent on the masses of the active sungrazers, which may vary widely over time depending on the size distribution of the LHB and NEA asteroid populations.

6.5   Results and Discussion

In the analytical model discussed here there are three sources of uncertainty. The first is simply the uncertainties arising in measured values taken from the literature, such as the chondritic and solar abundances of lithium and the total mass flux of the solar wind. The second source of uncertainty is in assumptions that must be made in the model. These include:

a) The amount of ionized dust component in the solar wind and the dust production rates.

b) The time evolution of the solar wind mass flux.

c) The amount of ^{6}Li (if any) produced in solar flares.

The third, and least constrained, source of uncertainty is in the assumptions in the impact rate, mass flux, and timescale of objects destabilized during the LHB. A very simple model of the mass flux of objects into the sun is adopted here in which contributions due to comets are ignored, the in-fall rate of asteroids follows a step-function increase followed by a logarthmic decay. The total mass of asteroids impacting the sun during the LHB is taken to be 2.4×10^{25} g, the post-LHB in-fall rate is taken to be 2×10^6 g s^{-1}, and the pre-LHB in-fall rate is 2×10^7 g s^{-1} (e.g. Levison et al., 2001; Bottke et al., 2002, 2005).

Due to these uncertainties, six different scenarios were investigated. The following is a summary of the assumptions in each case of the six cases:

- Case 1: The ionized dust component in solar wind has a mass flux equal to 1% of the asteroidal in-fall rate into the sun. This was chosen to match the current ionized dust mass flow rate from sungrazing comets found by Bzowski and Królikowska (2005). The solar wind mass flux is equal the current flux. No solar flare ^{6}Li production is considered.

- Case 2: The ionized dust component in solar wind has a mass flux equal to 10% of the asteroidal in-fall rate during the LHB, and 1% of the asteroidal in-fall rate all other times. The solar wind mass flux is equal to the current flux. No solar flare ^{6}Li production is considered.

- Case 3: The ionized dust component in solar wind has a mass flux equal to 10× the asteroidal in-fall rate. This was chosen so that the current solar wind ^{6}Li/^{7}Li ratio matched the value measured in lunar soils by **Chaussidon and Robert (1999)**. The solar wind mass flux is equal to the current flux. No solar flare ^{6}Li production is considered.

- Case 4: The ionized dust component in solar wind has a mass flux equal to 10% of the asteroidal in-fall rate during the LHB, and 1% of the asteroidal in-fall rate all other times. The solar wind mass flux follows the power law of **Wood et al. (2005)** after 400 My (given by Equation **6.3**). No solar flare ^{6}Li production is considered.

- Case 5: Ionized dust component in solar wind has a mass flux equal to 10% of the asteroidal in-fall rate during the LHB, and 1% of the asteroidal in-fall rate all other times. The solar wind mass flux is equal to the current flux. The solar flare ^{6}Li production rate has a constant value of 2.4 ± 0.7 g s^{-1} into the solar wind, chosen so that the current solar wind ^{6}Li/^{7}Li ratio matched the value measured in lunar soils by **Chaussidon and Robert (1999)**.

- Case 6: The ionized dust component in solar wind has a mass flux equal to 10% of the asteroidal in-fall rate during the LHB, and 1% of the asteroidal in-fall rate all other times. The solar wind mass flux follows the power law of **Wood et al. (2005)** after 400 My (given by Equation **6.3**). The solar flare ^{6}Li production rate has a constant value of 24 ± 7 g s^{-1} into the solar wind (10× the value in Case 5).

The ^{6}Li/^{7}Li ratio as a function of time between $100 - 1000$ My for each of the six cases is plotted in Fig **6.1**. The vertical width of the lines in each of the figures arises due uncertainties in the measured chondritic and photospheric abundances, uncertainties in the characteristic timescales for lithium depletion, and uncertainties the mass flux of the solar wind. A summary of results showing the peak ^{6}Li/^{7}Li ratio during the LHB, as well as the final isotopic ratio after 4.56 Gy is given in Table 6.1.

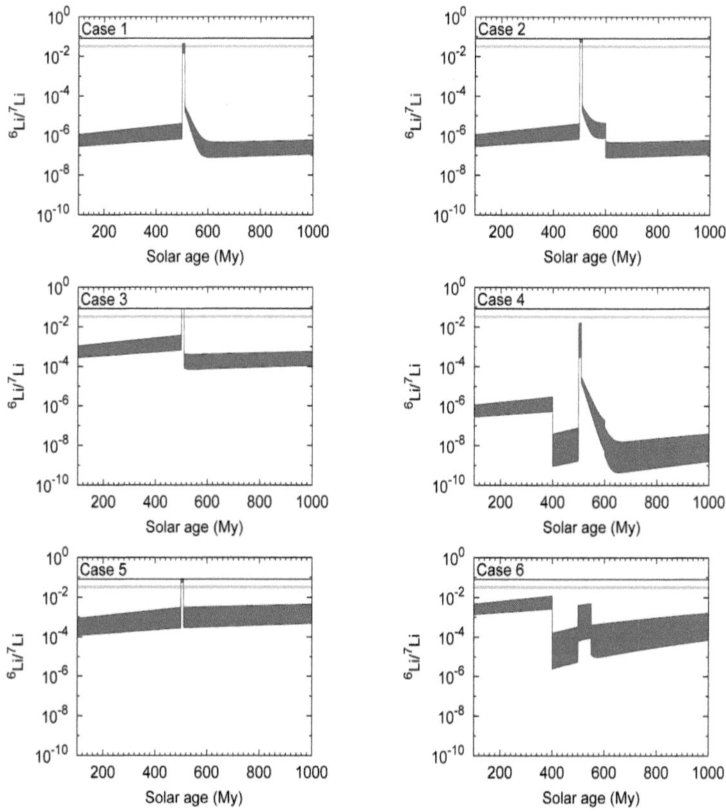

Figure 6.1: Solar wind lithium isotope ratios for six models of early solar system history

Case 1 is a fairly conservative case. It assumes that the only parameter that changes over time is the lithium abundance of the sun. There is a peak in the

Table 6.1: Lithium isotopic ratios during LHB and after 4.56 Gy for each model case considered.

Case	^{6}Li/^{7}Li — Peak LHB ($\times 10^{-2}$)	^{6}Li/^{7}Li — $t = 4.56$ Gy ($\times 10^{-2}$)
1	2.98 ± 1.56	0.0071 ± 0.0018
2	6.59 ± 1.03	0.0071 ± 0.0018
3	8.22 ± 0.2	0.647 ± 0.151
4	0.822 ± 0.818	0.0071 ± 0.0018
5	6.61 ± 1.04	4.51 ± 2.28
6	0.86 ± 0.83	45.1 ± 22.8

lithium isotope ratio due to the LHB, but it may not be measurable. Also, Case 1 is unable to reproduce the lithium isotope ratio measured in lunar soils at the present time (see Table 6.1).

In Case 2 the rate of ionized dust as a fraction of the asteroidal in-fall rate is assumed to be 10× higher during the LHB than either before or after. Since the dust production due to both collisions and the sublimation of sungrazing asteroids during the LHB is poorly constrained, this may be a reasonable assumption. This produces a spike in the lithium isotope ratio during the LHB that is comparable to ratio measured in lunar soils, so this is quite a promising result. However, without any other source of enhanced ^{6}Li during the present epoch, this model does not reproduce measured present solar wind lithium isotope ratio.

Case 3 was constructed to determine if ionized dust alone could account for the enhanced ^{6}Li measured in lunar soils. It was determined that the ionized dust component to the solar wind had to be 10× larger by mass than the total asteroidal in-fall rate into the sun. This is highly unlikely, otherwise the inner solar system dust bands would contain an order of magnitude or more mass than the entire Near Earth Asteroid population, and this is not observed. Nevertheless, the current major source of near solar dust may primarily be from the Kreutz family of sungrazing comets. If a much larger object became a sungrazing "comet," it could hypothetically produce a short-lived spike in the dust in-fall rate into the sun, and the lithium isotope ratio measured in the lunar soils may simply represent a sample from such period an enhanced dust-production period. Further work would need to be done to determine if the lunar soil samples analyzed by **Chaussidon and Robert** were well enough separated in exposure time to make such a scenario unlikely.

In Case 4, the promising results of Case 2 were revisited, but this time with an enhanced solar wind mass flux given by Equation **6.3**. These results illustrate how the ionized dust component is perhaps the most important parameter in this study. Despite an identical asteroidal in-fall rate, the final solar wind ^{6}Li/^{7}Li is an order of magnitude lower than when the solar wind mass flux was equal to its modern value, as in Case 2. The dilution of the ionized dust by the enhanced solar wind may make detection of the chondritic lithium spike at the LHB difficult.

In Case 5 the creation of ^{6}Li in solar flares was included. If the production rate of the lighter lithium isotope remained constant throughout time, this produces a measurable signature in the lithium isotope ratio during the LHB as well as reproducing the measured isotope ratio in modern lunar soils. The assumption that ^{6}Li production is constant throughout time may not be a good one, however. If the production rate of the lighter isotope was related to the frequency of solar flares, a younger more active Sun during the LHB may have produced more ^{6}Li than is currently seen.

Finally in Case 6 both the ^{6}Li isotope production and solar wind mass flux were increased. This was an attempt to model a significantly more active Sun during the LHB. The ^{6}Li production rate was increased by $10\times$ over the current value prior to a solar age of 1 Gy. There is still a spike in the lithium isotope ratio, but it is not well separated outside of the uncertainties and it may be below the threshold of measurability.

6.6   Conclusion

The results above are promising but inconclusive. Better constraints on the inputs are needed in order to determine whether or not a distinct signature of the LHB is preserved in the lunar record. If the Sun were significantly more active at the time of the LHB, such as in Case 4 and 6, then evidence of the LHB in implanted solar wind particles may be difficult to detect.

The most important parameter in the analytical model developed here is the fraction of interplanetary dust that is ionized and injected directly into the solar wind. This is because the density of the solar wind is much lower than mean density of the solar convective zone, therefore chondritic material in the form dust that is converted to solar wind pickup ions has a greater effect on the solar wind lithium isotope ratio than material that impacts the Sun, despite being much less massive. Ionized dust alone is unlikely to explain the enhanced ^{6}Li/^{7}Li ratio measured in lunar soils. A time history of solar wind lithium may be preserved in other lunar samples, such as breccias that contain soil grains that were exposed to the solar wind in the past but were later protected.

Future work in this area should include developing a more realistic model of the in-fall rate of asteroidal and cometary material, including production, sublimation, and ionization rate of near-solar dust. A better constraint on the solar wind mass flux and the solar lithium abundance and destruction rate near the time of the LHB is also desired.

CHAPTER 7

SUMMARY AND CONCLUSIONS

I have explored the dynamical history of the asteroid belt using dynamical models of the main asteroid belt. Included here is a short summary of my conclusions, followed by a brief discussion about implications and possible future areas of study.

I find that the present-day distribution of large asteroids in the main belt is consistent with the effects of planetesimal-driven giant planet migration. I have placed constraints on the rate of migration of the giant planets using an analytical model of the effects of the sweeping ν_6 secular resonance on the inner asteroid belt. The constraint that the migration rate of Saturn must have been > 0.2 AU My^{-1} is much faster than rates of migration of the giant planets found in the literature. Reconciling these differences presents a major challenge in our understanding of this important period of the solar system's history.

The asteroid belt has experienced a significant amount of depletion due to dynamical erosion since the epoch when the current dynamical architecture of the main asteroid belt and the major planets was established. This dynamical erosion affects asteroids of all sizes, and is the dominant loss mechanism for asteroids with $D > 30$ km in diameter. The best model for describing the loss history as a function of time is $\dot{n} \propto t^{-1}$, where the proportionality constant undergoes changes at finite periods of time. The asteroid belt at $t = 200$ My (where $t = 0$ is when the asteroid belt reached its current dynamical state) would have had 28% more large asteroids than today, and the asteroid belt at $t = 10$ My would have had 64% more large asteroids than today. Extending this model to asteroids with diameters $D > 10$ km and using models of the estimated impact probabilities onto the terrestrial planets, I find that the impact flux of large craters (craters with final diameters $D_c > 140$–200 km) is an order of magnitude lower than estimates used in impact crater chronologies and impact risk assessments.

The solar photosphere is currently highly depleted in Li relative to chondrites, and the ^{6}Li is expected to be far less abundant in the sun than ^{7}Li due to the different nuclear reaction rates of the two isotopes. The addition of meteoroids into the Sun during the giant planet migration may have driven the solar wind ^{6}Li/^{7}Li ratio toward chondritic values. In addition, sublimation of asteroids on high eccentricity "sungrazing" orbits and the collisions between objects in the inner solar system may have produced dust, some of which could become ionized and injected directly into the solar wind as pick-up ions with chondritic lithium isotope abundances. I have attempted to quantify the change to the solar wind ^{6}Li/^{7}Li ratio due to the estimated in-fall of chondritic material and enhanced dust production during planetesimal-driven giant planet migration. Evidence for a short-lived impact cataclysm that affected the entire inner solar system may be found in the composition of implanted solar wind particles in lunar regolith, thereby helping determine the timing and duration of this event. However, much remains to be understood about the interaction of meteoritic material and the solar atmosphere and my results are inconclusive. Understanding the production of dust during planet migration (both due to comminution of planetesimals and solar sublimation) is key to understanding how the solar wind lithium abundance may have changed due to the in-fall of chondritic material.

7.1 Further exploration of the effects of planet migration on the main asteroid belt

In Chapter 3 I demonstrated that the observed distribution of large asteroids (where absolute visual magnitude is used as a proxy for size) is consistent with the effects of planetesimal-driven giant planet migration. However, there are a number of ways in which my model could be improved. I performed my planet migration simulation on a model asteroid belt that had already experienced 4 Gy of dynamical evolution. An obvious improvement would be to put the models into the "correct" order, that is do the planet migration simulation first, followed by the 4 Gy long dynamical erosion simulation described in Chapter 2. Paradoxically however, performing the simulations in the correct order would introduce new complications. As I demonstrated in Chapter 5, the eccentricity distribution of the asteroid belt after planet migration is a function of both the rate of migration of the planets and the pre-migration eccentricity distribution. It is unclear whether the present-day distribution of asteroid eccentricities (see Fig. 5.5a) contains enough information to determine the pre-migration distribution. A constraint on the pre-migration state of the asteroid belt may need to come from studies of the primordial excitation of the asteroid belt, such as in O'Brien et al. (2007).

A better model of the effects of planet migration would also ideally contain roughly 10–20 times as many particles as were used in the Chapter 3 model asteroid belt. This requirement motivated the development of Sim 2 described in Chapter 2, however that simulation was very computationally expensive, and required several months of computing time in order to complete ~ 1 Gy of integration time. Performing the simulations in the correct order would help, as most of the initial number of asteroids should be lost after the initial phase of planet migration. Another advantage of doing the simulations in the correct order would be an investigation into the capture of the Hilda population in the 3:2 jovian mean motion resonance, similar to the study done by Franklin et al. (2004).

I have also neglected the important pre-migration history of the main asteroid belt. If, as some evidence suggests, sweeping resonances in the asteroid belt due to giant planet migration is the cause of the LHB at 3.9 Gy ago (e.g. Strom et al., 2005), then the asteroid belt experienced ~ 700 My of collisional and dynamical evolution before planet migration began. Such a long stretch of time should have been enough to sculpt "primordial Kirkwood gaps" at the earlier locations of jovian mean motion resonances. Especially important is the location of the ν_6 resonance at this early times. As I demonstrated in Chapter 3, the distribution of the large asteroids in the main belt is consistent with the effects of resonance sweeping due to planet migration, but there do not appear to be regions of excess depletion due to long-lived resonances anywhere in the main belt. There are several possible solutions to this dilemma. One is that the pre-migration eccentricities of the giant planets was much lower. This is consistent with the Nice model, which proposes that the current eccentricities (and inclinations) of the giant planets were produced during the violent instability arising from the crossing of the Jupiter-Saturn 2:1 resonance (**Tsiganis et al., 2005**). In this model, the giant planets had more circular orbits in the 700 My period before the resonance crossing initiated rapid planet migration. The strength of the ν_6 resonance is related to the power of the g_6 mode in the giant planets, which is related to the eccentricity of the giant planets. As Eqs. (**5.31**) and (**5.3**) illustrate, the strength of the resonance (given by the coefficient σ) is proportional to the square of the power of the g_6 mode in the planets. Therefore, the ability of the ν_6 resonance in clearing asteroids is very sensitive on the eccentricity of the giant planets. Alternatively, outer solar system planetesimals that were destabilized by planet migration and then captured by sweeping jovian mean motion resonances could potentially have filled up any ancient gaps (**Levison et al., 2009**).

Another possibility is that the epoch of planet migration occurred very soon after the planets formed, and that the LHB was caused by some other agent, such as a quasi-stable planetary embryo in the "Planet V" hypothesis (**Chambers, 2007**). This is difficult to reconcile with various pieces of evidence that link the LHB to the main asteroid belt. Evidence in the size distribution of cratered landscapes of the Moon, Mercury, and Mars that are associated with the LHB show the signature of main belt asteroids, implying some type of size-independent (i.e. dynamical) perturbation on the main asteroid belt (**Strom et al., 2005**). In addition geochemical evidence points to an asteroidal source for lunar impactors during the LHB (**Kring and Cohen, 2002**). As discussed in Chapter **3**, the distribution of asteroids in the main belt is consistent with the effects of giant planet migration. At least $\sim 10^{21}$ g of material is required to have impacted the Moon during the LHB, based on the observed abundance of ancient lunar craters (**Chyba, 1990**). Based on the impact probabilities calculated in Chapter **2** and the mass of the present-day asteroid belt, the LHB requires a minimum of ~ 2 times the present-day mass of the asteroid belt to have been in terrestrial planet-crossing orbits. Any perturbation to the asteroid belt large enough to produce the LHB should have left its mark on the structure of the main asteroid belt. If planet migration occurred long before the LHB, then an either an additional perturbation on the asteroid belt that somehow preserved the earlier signature of planet migration in the distribution of asteroids occurred, or some other reservoir of small bodies that happened to have an identical size distribution as the main asteroid belt (and similar geochemistry) had to have been destabilized.

7.2   Secular theory for a migrating planetary system

In Chapter 4 I calculated the secular dynamics of the solar system during planet migration. A first order Laplace-Lagrange secular theory with corrections for the near 2:1 Jupiter-Saturn mean motion resonance, similar to that derived by **Malhotra et al. (1989)** for the uranian satellite system, is adequate to describe the behavior of the secular frequencies over a large range in Saturn's semimajor axis. The exception to this is in the current solar system, where the effects of the near 5:2 resonance, the "Great Inequality," increase the secular frequencies by a substantial amount and introduce additional frequencies. When Jupiter and Saturn are within a resonance, such as the 5:2, 7:3, 2:1, and others, the e-ϖ secular frequencies broaden and the power spectra become noisy. In such cases it can be difficult to identify the equivalent g_5 and g_6 frequencies, and, especially near the 2:1 resonance, the g_5 mode dominates over the g_6 in Saturn's e-ϖ time history. The strength of the ν_6 resonance is related to the power of the g_6 mode in the planets' secular evolution, and in power spectra analysis the power of a mode is proportional to the amplitude of the peak at that mode. Therefore, if Saturn and Jupiter were once closer to the 2:1 resonance than today the ν_6 may not have been as powerful as today. A better understanding of the secular dynamics near Jupiter-Saturn mean motion resonances may help refine models of planet migration and better understand the effects of secular resonance sweeping on the asteroid belt. In addition, I have ignored the important coupling between asteroid eccentricities and inclinations during the calculation of the position of the ν_6 resonance. The inclination dependence of the ν_6 resonance location is well known for the current solar system, but not for the solar system with the giant planets undergoing migration.

7.3   Improving the terrestrial planet impact flux calculation

In Chapter 2 I estimated the rate of $D > 10$ km asteroid impacts onto the Earth and Moon using my dynamical model of the asteroid belt. I found that my estimate was lower by roughly an order of magnitude than to estimates in the literature of the production rate of $D_c > 140$–200 km craters. However, my model did not incorporate collisional evolution or non-gravitational effects. An asteroid belt model containing both dynamical evolution as well as collisional evolution over 4 Gy of solar system history, and incorporating the effects of non-gravitational forces such as the Yarkovsky effect, could provide deeper insights into the history of the asteroid belt and its contribution to the collisional history of the terrestrial planets. Whether or not additional model parameters can account for a factor of ~ 10 discrepancy in the number of $D > 10$ km asteroid impacts is uncertain. However, the result that the large ($D \gtrsim 30$ km) asteroid population has significantly eroded over the age of the solar system is robust.

Because the large asteroid population is the reservoir from where collisional fragments that become NEAs are derived, then if the reservoir has eroded then the rate of production of small fragments should have declined as well. This may provide an explanation for the result of **Quantin et al. (2007)** that suggest that the production of $D_c > 1$ km craters has declined by a factor of three over the past ~ 3 Gy. This hypothesis has not been explored in this work, and a more complete model of the asteroid belt incorporating dynamical, collisional, and non-gravitational effects would be needed to definitively test it. This hypothesis also needs not contradict the evidence that the impact flux in the terrestrial planet region has increased by a factor of ~ 2–3 in the past ~ 500 My (**Marchi et al., 2009**). The impact cratering rate on the terrestrial planets may have undergone significant fluctuations on top of a long-term decline, due to breakup events in the asteroid belt (e.g., **Korochantseva et al., 2007; Zellner et al., 2009; Bottke et al., 2007b**).

Better models of the asteroid belt could help refine impact crater chronologies for the inner solar system.

7.4   Meteoritic pollution of the Sun and other stars

In Chapter 6 I explored the effect of in-falling meteoritic material in the solar atmosphere, specifically looking at the element lithium and its isotopes. Extending this study to include observations of young sun-like stars could help determine whether or not planetesimal-driven planet migration is a common feature of the early evolution of planetary systems. Many sun-like stars appear to have observable transient warm dust disks that could be analogues to the dust produced by scattered planetesimals during our solar system's planet migration (Wyatt et al., 2007). If planet migration is a common feature of planetary systems, then the signature of meteoritic infall may be observable in the lithium abundances of stellar atmospheres where planetesimal-driven migration is occurring.

7.5   Parting thoughts

The epoch of planetesimal-driven giant planet migration was a unique period when events far out in the distant realm of Pluto shaped the history of every planet from massive Jupiter down to tiny Mercury. The tentative links between this period of time and the Late Heavy Bombardment that left its scars in the cratered landscapes of the planets and satellites hint at a profound moment of violence in the early history of Earth. There is much yet to be learned about this fascinating period of our solar system's history. I end this work with a bit of whimsical speculative poetry inspired by this event.

Bartholomew Lombardment and the Late Heavy Bombardment

A long time ago, four billion years give or take
Lived a microbe named Floyd in a cozy little lake

On a nice sunny day Floyd was highly annoyed
When his home was destroyed by a big asteroid

He shlorped and he flooped and got quite red in the face
As his single-celled body was flung deep into space

"What is this? What is happening!?" he cried out with fear
But no one could answer, there was nobody near!

One day there arrived a man in a hat
He pulled out a briefcase and gave Floyd a small pat

"I'm a lawyer! My name is Bartholomew Lombardment
And I represent the Late Heavy Bombardment"

"A lawyer!? A lawyer!?" cried Floyd, quite perturbed.
"I don't need a lawyer!
I need a house with a fireplace and foyer!
I've been kicked to the stars!
I've been sent way too fars!
I want to go back to my lake down on Mars!"

"There's no need to fret, there's no need to fear"
Said Bart as he showed him a chart that was near.

"It's called the Late Heavy Bombardment, you see,
I'll explain it all to you," he said with much glee.

"Jupiter and Saturn are restless, their orbits are changing
With a cold disk of iceballs, their momentum's exchanging

Now the asteroids can't all remain where they are
Perturbations from Jupiter will fling them afar.

They'll crash down on Mercury

They'll crash down with fury!

They'll crash down on Mars
Like big shooting stars!

They'll crash down on Venus, the Earth, and the Moon,
But don't worry so much it will all end quite soon!"

A hundred million years went by
Of rocks falling from the sky

When Floyd finally landed, his heart filled with mirth
While he made his new home right down here on the Earth.